Environmental Management for Hotels

DATE DUE

Sept 14			
Mas			

Environmental Management for Hotels

A student's handbook

David Kirk

)X2 8DP

A member of the Reed Elsevier plc group

OXFORD   LONDON   BOSTON
MUNICH   NEW DELHI   SINGAPORE   SYDNEY
TOKYO   TORONTO   WELLINGTON

First published 1996

British Library Cataloguing in Publication Data
Kirk, David
   Environmental Management for Hotels:
   Students Handbook
   I. Title
   647.940682

ISBN 0 7506 2380 2

Typeset by David Gregson Associates, Beccles, Suffolk
Printed and bound in Great Britain by Scotprint Ltd., Musselburgh

Contents

Preface   vii

Acknowledgements   ix

1 **Introduction**   1
What do we mean by the environment? – The driving forces for
change – Sustainability and the protection of scarce resources –
Global environmental issues – Tourism, hospitality and the
environment – References and further reading

2 **Environmental management**   16
The environmental system – Environmental policy, strategy
and implementation – Environmental impact assessment –
Case studies – References and further reading

3 **Water management**   32
Water and the environment – Water supplies – Improving water
quality – Control of water consumption – Case studies –
References and further reading

4 **Energy management**   47
The principles of energy management – Energy supplies – The
energy management programme – Case studies – References and
further reading

5 **Management of the indoor environment**   80
The significance of the indoor environment – Chemical hazards –
Air quality – Noise – Light – Non-ionizing radiation –
Case studies – References and further reading

6 **Materials and waste management**   102
The need for materials and waste management – The waste audit
– Product purchasing – Operations management – Environmental
pollution – Recycling – Case studies – Summary – References and
further reading

Index   127

Preface

I was present at the launch of the International Hotels Environment Initiative (IHEI) on 31 May 1993. I was very impressed with the guide produced by the IHEI and I immediately saw its potential to students and as a teaching aid to lecturers in a wide range of areas of the hospitality curriculum, such as accommodation management, facilities management and hospitality operations management.

However, given the cost and format of the guide, I suggested to staff of Butterworth-Heinemann present at the launch that it might be worth considering a student edition, which would contain more in the way of background theory and explanation and less operational detail. They agreed to put the idea to the IHEI who also thought the idea worth pursuing.

The text of this book is aimed primarily at students on postgraduate, undergraduate and HND courses in hotel, catering and hospitality management. It should also be suitable for students on vocational hospitality courses who are involved in project work on environmental management. Early in the process of writing the text, I took the decision to retain the focus on hotels rather than develop the book into the more general area of hospitality. Whilst it could be argued that it would be more useful if the book covered a broader range of operations which constitute the hospitality industry, I feel that it is better to retain a focus on a single type of operation, allowing a holistic approach which emphasises the interactions which take place. In my view, it should be relatively easy to then apply these principles to other areas of hospitality.

In this book I have attempted to relate environmental management to the general management of hotels since, the concepts are most likely to be accepted if they are integrated into the overall framework of decision making and day-to-day management. Environmental management cannot succeed if it is seen as an 'add-on' to the management decision making process. I have retained the case studies which were developed in the guide as these allow students to see the links between the general principles developed in the text of the book and the way in which industry has chosen to develop these principles.

The book starts with a general introduction to the concept of sustainability and develops the idea that we need to take action locally if we wish to change the global environment. This is followed by a discussion of some of the major threats to the environment and their causes. Specific environmental initiatives within the hospitality and tourism industry are then described. In Chapter 2, the underlying principles of environmental management are developed through agreed policies, an audit of current practice and the targeting of areas which would benefit from change. In doing this, emphasis is placed on the need to develop environmental awareness throughout the company, all the way from board level down to all levels of staffing, and to identify individuals who will take responsibility for action.

The two chapters which follow go on to look at two of the major areas of resource consumption in a hotel, water and energy. Some hotels have already done much to reduce consumption and associated costs in these areas and these projects are illustrated through the case studies. Because of the high cost of energy and water to hotels, it has been possible to institute changes both at an operational level (through awareness and training) and through capital projects. Chapter 5 goes on to look at the management of the quality of environment within the building, with specific reference to air quality, noise and lighting.

The final chapter looks at the management of materials and waste, taking a holistic view of materials management from purchasing to waste disposal, with a discussion of the relative merits of waste elimination, waste re-use, recycling, incineration and land-fill disposal. This is followed by a summary which considers the relationships between all of the undesirable outputs from the hotel and indicates how they should be viewed as a total management system rather than as separate problems.

David Kirk

Acknowledgements

Thanks must go to the International Hotels Environment Initiative for their agreement to develop a student version of their excellent manual *Environmental Management for Hotels: The Industry Guide to Best Practice* and for allowing the use of text, diagrams and case studies from this manual. In addition to the use of work from this guide, Figures 4.13, 4.14, 4.15 and 4.16 are taken from *Kitchen Planning and Management* by John Fuller and David Kirk, published by Butterworth-Heinemann (1991).

Thanks must go also to my wife Helen for her patience and her willingness to allow the lap-top computer to accompany us everywhere, including on holidays to the beautiful Lake District, which provided a sufficient inspiration to finish the book.

1 Introduction

What do we mean by 'the environment'?

While there are several different ways in which the word 'environment' is used, most people are aware that there is a need for all of us to take care of the environment, if we are not to threaten the ability of the earth to support future generations. Some aspects of the environment are very obvious from our day-to-day lives, such as increasing traffic levels, together with the associated air pollution and loss of greenbelt (protected areas of land surrounding towns and cities) and the countryside to road development and urbanization. We are aware of other dangers through the debate in the media, but these issues vary from tangible effects such as the shortage of physical resources (such as fossil fuels) to less evident and more long-term effects such as global warming and the hole in the ozone layer. The difficulty lies in translating these overall concerns, particularly those that are not directly related to us and which are less tangible, into action by the organization and by the individual. Our actions can sometimes seem inconsequential, compared to the size of global problems (Wright, 1992).

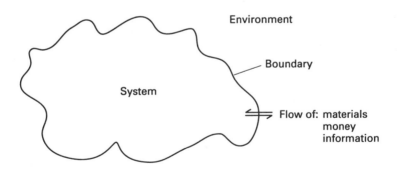

Figure 1.1 *A system and its environment*

From a systems viewpoint, the term 'the environment' refers to any aspects that lie outside the system under consideration and which are separated from the system by a boundary (see Figure 1.1). The boundary acts as a control on the flows that take place from the system into the environment and vice versa. In this context we can consider the boundary to be a semi-permeable membrane which acts as a regulator, allowing the free flow of some things but preventing the flow of others (see Figure 1.2).

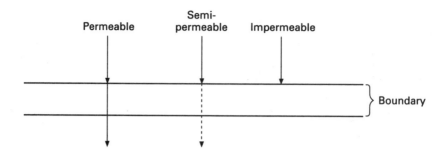

Figure 1.2   *Control of flows across a boundary*

The boundary can be considered to be an artificial construct which allows us to define our area of interest. For example, we can think of our environment in terms of global, continental, national, regional, local and personal boundaries. In this way, systems form hierarchies. At one level, we can think of a person as a system, but then people working together in a hotel make up the human resource system. In a different context, these same people form the local and national social and political systems. Beyond this, individuals represent their national social and political groupings at international gatherings.

We can view some of the environmental issues, particularly the relationship between our actions and the environmental impacts in terms of primary, secondary and tertiary effects, as shown in Figure 1.3. At a local level, we might decide to scrap a number of old refrigerators which contain chlorofluorocarbons (CFCs). We sell the old refrigerators to a local scrap metal merchant who crushes the refrigerators so that the metal can be sent to reprocessors, which results in CFCs being released into the atmosphere, but the amount of CFC is very small compared to the total emission of these gases. The release of CFCs is thought to contribute to the development of holes in the ozone layer, a secondary effect of the release of CFCs. These holes are thought to have a number of effects, such as causing an increase in cases of skin cancer caused by increased levels of ultra-violet radiation – a tertiary effect of the release of CFCs.

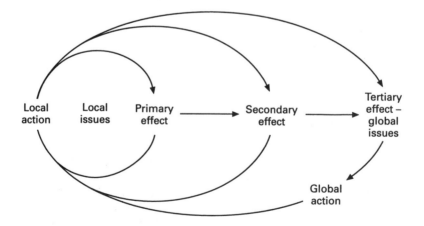

Figure 1.3   *Think globally, act locally*

Another example might be an increased use of electricity through the installation of air-conditioning, resulting in the need for more fossil fuel to be burned at a power station, causing increased emissions of carbon dioxide (CO_2) and sulphur dioxide (SO_2). These in turn cause acid rainfall over countryside at a great distance from the power station and the acid rain increases the acidity of lakes, killing the flora and fauna. We need to be able to make these links between local action and the secondary and tertiary global effects so that we can modify our local actions and halt some of these changes. If we are aware of these chains of cause and effect we can change our actions and convert some of these vicious circles of cause and effect into virtuous circles, where our actions minimize negative impacts and maximize positive ones.

Discussions take place, decisions are made and actions taken at global, national and local levels, and may result in a number of different outcomes, such as:

- International agreements
- National and international laws
- National/local policies
- National/local pressure groups
- Company policies and actions
- Individual actions.

Environmental problems must be tackled at all these levels. There is a clear need for global policy making and target setting, such as the Montreal Protocol of 1987, which established targets for CFC emissions. Another example might be the United Nations Rio Earth Summit Conference in 1992, at which a number of developed countries agreed that, by 2000, they would reduce the level of carbon dioxide emissions in their countries to the levels of 1990. The European Union has introduced several Directives which relate to the management of the environment. Within the UK, there has been a long history of legislation related to the protection of the environment, including the Clean Air Act 1956 and the Control of Pollution Act 1974. Much of this early legislation was not directly related to the management of the environment in a holistic sense but was concerned with preventing gross pollution, largely related to health issues. However, as a response to the global issues of the Brundtland Report, the British government has produced a White Paper on the environment (HMSO, 1990).

Agreed policies by themselves will not necessarily cause people to change their habits. Action is required at a national and local level if any real changes are to result. Therefore we need to look at the possible driving forces to change at a local level, which are essential if any of these global issues are to be addressed.

The driving forces for change

There are five main forces for change within an organization:

1 Legislative and fiscal requirements
2 Advantages resulting from financial savings
3 Consumer attitudes
4 Public opinion
5 Enlightened management.

In relation to the last point, some companies recognize the importance to the company of its social and environmental responsibilities. Companies are now being measured not only on their financial performance but also on their ethical performance. This affects both shareholders and consumers and a number of investors take a great deal of interest in the broad range of ethical issues facing a company, including the environment. However, it is often difficult, particularly for the small organizations, to know how to respond to these issues and to generate the resources needed to do so.

Over recent years companies have become aware of the development of pressure groups and green politics. Steven Young, in his book on *The Politics of the Environment*, reviews the development of environmental pressure groups and green political parties (Young, 1993). Many of the early groups, going back to the Commons, Open Spaces and Footpaths Preservation Society (established in 1865), were concerned with the natural heritage together with the flora and fauna. Groups with a broader political agenda started much later with, for example, Friends of the Earth in 1971 and Greenpeace in 1977. From being very radical, many of these groups have now become a much more central part of pressure politics and operate through tactical alliances on specific issues. Green political parties started in the 1970s, partly as a result of many of the pressure groups such as Friends of the Earth. They achieved the greatest success in Europe, particularly in Germany and Switzerland. While green political parties have had mixed electoral success, they have had a distinct impact on more conventional political parties and governments.

Environmental awareness among consumers has increased, albeit from a low base until now most manufacturers have responded even if only in a minor way to these concerns and a number of manufacturers have developed specific eco-friendly products.

Sustainability and the protection of scarce resources

Sustainabilty

One of the most fundamental principles of environmental management relates to the establishment of sustainable development. The Brundtland Report (World Commission on Environmental Development, 1987) defined sustainable development as 'development that meets the needs of the present without compromising the ability of future generations to meet their own needs'. This can be viewed in a number of ways which relate to physical and social factors. In physical terms, there is a need to conserve scarce physical resources such as minerals and to minimize negative impacts on the physical environment through pollution. Any buildings or other form of built development needs to be sensitive to social, political, cultural and geographical aspects of the site chosen for development.

The concept of sustainable development has been expanded to cover seven key aspects (Young, 1993):

1 Futurity: developments must be considered against a longer time-span than that normally used by businesses and politicians.
2 Inter-generation equality: current activities should not deplete the resource base available to future generations, so that a constant resource capital can be passed on.

3   Participation: all political and social groups affected by a development should be involved in debate and decision making.

4   The balancing of economic and environmental factors: decisions should be made on the basis of a broader range of issues than the economic costs and environmental issues should be elevated from that of a constraint on development.

5   Environmental capacities: all environmental impacts should be assessed in terms of their effect on equilibrium processes so that delicate ecological balances are not disturbed.

6   Emphasis on quality as well as quantity: decisions should not be made on the basis of 'least-cost' but on a solution which gives the least damaging long-term solution.

7   Compatibility with local ecosystems: to ensure that developments sustain local social, political, agricultural and ecological systems.

In reviewing the effect of a business on sustainability we can view its operations in terms of its inputs (the way in which it consumes resources) and its outputs (the negative and positive impacts on the environment).

Inputs: renewable and non-renewable resources

Although this is a simplification, it is possible to consider resources as either renewable or non-renewable. A non-renewable resource is one for which there is a finite supply which, once it has been used up, cannot be replaced. The easiest way to think of this is that there is a store of the material from which we can draw supplies, thereby reducing the stock (see Figure 1.4). The amount of material remaining depends upon the rate of use. Examples of non-renewable resources are minerals such as copper and lead. Other examples include the fossil fuels. The fossil fuels are a good example; they resulted from the conversion of the energy of the sun into chemical energy, in the form of plant and animal tissues over many millions of

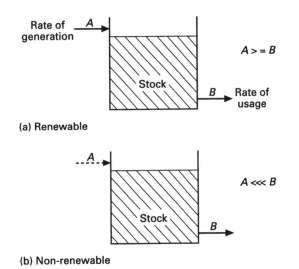

(a) Renewable

(b) Non-renewable

Figure 1.4   *Renewable and non-renewable resources*

years. Although, in principle, the conditions which led to the development of coal, gas and oil still exist in the world, the rate of extraction is so much greater than the rate of formation that they are, for all intents and purposes, non-renewable. It may be argued that, as materials become more expensive, there is a greater incentive to seek out and exploit new supplies. To some extent this is true, but all this does is to extend the life of the supply; it is still finite. In order to conserve supplies, the best approach is to reduce the rate of usage of these materials and to recover as much of the resource as possible after use.

In contrast, renewable resources are constantly being replenished and the rate of use does not affect future availability. Very few resources are fully renewable under all conditions but a good example of one is solar energy. No matter how much solar energy we use for growing crops, heating buildings or generating electricity, this does not affect the rate of supply from the sun. However, solar energy is an exception and with most renewable materials we are concerned with maintaining the balance between the rate of production and the rate of use. If we use a resource at a higher rate than that at which it is being regenerated, then we start to consume stocks and the supply declines. The material is only renewable up to the point where we have an equilibrium between the rate of regeneration and the rate of use. For example, there is a natural rate of regeneration of fish in the seas. Provided that we have sensible policies to control the amount of fishing and the protection of young, immature fish, future supplies can be protected. If we overfish and if we catch immature fish, then we upset the equilibrium and fish stocks diminish.

If the demand becomes greater than the rate of supply, it is sometimes possible to expand the rate of supply. For example, it has been possible to increase the yield of agricultural land by the use of new varieties of crops and by increasing the level of fertilizer. These benefits have, in many cases, been found to be short-lived as the increased use of fertilizer works on the basis of a law of diminishing returns and loses its effectiveness and also damages the soil. An additional factor is that monoculture increases the risk of plant diseases and pests.

Disposal of unwanted materials into the environment

For much of our history it had been assumed that the environment was so large that it could act as a huge rubbish-bin for all our unwanted material. It was also assumed that it would be possible to 'dilute and disperse' these wastes into the environment. The environment would either, at best, process these wastes into something non-harmful or, at worst, was sufficiently large that any harmful materials would be diluted to a level at which they were effectively dispersed.

It is now understood that this is not the case. The ability of the atmosphere, rivers, seas and land to absorb our waste with no detrimental effect is limited and, beyond that limit, any additional waste will have a damaging effect on the environment. Also, because of the complexity of ecological systems, the effects are not always predictable. For example, pesticides, developed to control plant diseases, are now found in significant levels throughout the food chain and not just in plants but also in animals. It is also interesting to note that CFCs were developed as a less hazardous alternative to other toxic or inflammable refrigerants. They were considered to be hazard free because of their inert chemical properties. This very inertness is now contributing to the destruction of the ozone layer.

Global environmental issues

The atmosphere

The atmosphere surrounding the planet earth is made up of two main layers: the troposphere (0–15 km), which is the layer in direct contact with the ground, and, above that, the stratosphere (15–50 km). Within the atmosphere there is a complex balance of energy and chemical composition taking place. It the past it was considered that the atmosphere was so large that any materials discharged into it would be diluted to such an extent that there would be no change in its composition. However, recent scientific evidence indicates that this is not the case and that the composition of the atmosphere is changing as a result of the effect of human activity. Some of these changes occur close to the earth's surface and are highly visible, such as fogs and smogs caused by burning fossil fuels. Other effects take place at much higher levels and are not visible. There are three major effects.

Global warming

Within the atmosphere, there is a complex energy balance taking place (Harrison, 1992, pp. 8–11), with incoming solar radiation being distributed and dispersed in a number of ways:

- Reflected from the upper surface of the atmosphere
- Absorbed by the atmosphere
- Reflected from the surface of the earth back into space
- Circulated in the atmosphere by convection currents
- Absorbed by the surface of the earth.

The atmosphere retains heat at the earth's surface and, without it, the surface temperature would be well below freezing point instead of the global average of 15°C. The lower layers of the atmosphere are transparent to visible light and the energy in this radiation is absorbed by the earth's surface and some of it is radiated back into the atmosphere. However, because of the low temperature of the surface of the earth

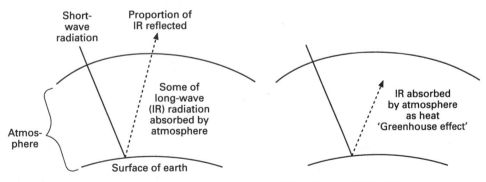

(a) Normal levels of CO_2

(b) High levels of CO_2, H_2O, etc.

Figure 1.5 *The greenhouse effect*

relative to the sun, this radiation is at a much lower temperature and is emitted as long-wave infra-red radiation. Many of the gases in the atmosphere are not transparent to this radiation and, because of this, the energy content of the radiation is absorbed into the atmosphere, causing a rise in temperature. This is often referred to as the 'greenhouse effect', as shown in Figure 1.5.

There are a number of greenhouse gases, so called because they are absorbers of long-wave radiation, including carbon dioxide, methane, moisture vapour, nitrogen oxides and CFCs. Carbon dioxide is the most abundant of these gases, but some of the less abundant gases are very important because of their relative absorbency of long-wave radiation (see Table 1.1)

Table 1.1   Relative absorbency of long-wave radiation

Gas		Relative absorbency $(CO_2 = 1)$
Carbon dioxide	CO_2	1
Methane	CH_4	20
Nitrous oxide	N_2O	200
Freon 11 (R11)	$CFCl_3$	12,000
Freon 12 (R12)	CF_2Cl_2	16,000

One of the most significant effects of human activity has been a gradual shift in the proportion of carbon dioxide in the atmosphere (Goudie, 1993). Since the start of the Industrial Revolution, there has been a steady increase in the proportion of carbon dioxide in the atmosphere, caused by the combustion of fossil fuels. Deforestation and the associated burning of the timber have also contributed to the build-up of carbon dioxide. Since all plants convert carbon dioxide into oxygen, the deforestation process has had this double effect. The result of all these processes is that the level of carbon dioxide in the atmosphere is increasing by about 2 per cent a year, from a level of about 260 parts per million (ppm) in pre-industrial times, to 300 ppm at the turn of the century and now is over 350 ppm. It is predicted that, by the middle of the next century, it will have risen to over 450 ppm. In addition, many of the other greenhouse gases, such as methane, nitrous oxide, krypton 85, water vapour and the CFCs, are increasing at a similar rate.

Humidity in the atmosphere also plays an important part in controlling temperature, as we can see from the way in which humid regions of the world stay warm at night compared to arid deserts. A resultant effect of an increase in temperature is that there may be a greater rate of evaporation from seas and lakes, causing an increase in humidity. This increasing humidity in the atmosphere is expected by some scientists to contribute to the greenhouse effect, because water vapour also absorbs long-wave radiation.

It is difficult to predict the precise effect on the climate of this increase in the greenhouse gases, especially as the climate fluctuates by quite large amounts due to natural factors unrelated to human activities. It may take a long time before the effect of the increase in greenhouse gases on climate can be proved but, in the meantime, there is considerable international effort to reduce the emission of these materials, particularly carbon dioxide and CFCs. It is forecast that the temperature may rise by 1°C every 30 years as a result of the greenhouse effect.

The effects of global warming are predicted to be higher temperatures, increased rainfall and storms, rises in sea level (causing flooding of low-lying areas) and increased algal growth in coastal areas.

Ozone depletion

The ozone layer, which occurs 25 km above the earth's surface in the stratosphere, plays an important role in absorbing ultra-violet radiation (particularly the harmful UV-B) from the sun and preventing most of it from reaching the surface of the earth. Since ultra-violet radiation is harmful, resulting in the short-term effect of severe sunburn and the longer term effect of skin-cancer, the ozone layer plays a very significant role. Any reduction in the ozone layer would lead to an increase in the amount of ultra-violet radiation and increasing risk of sunburn and skin cancer. Ozone occurs in equilibrium in the stratosphere, but this equilibrium is being disturbed by a number of human activities.

Some gases have been shown to be responsible for displacing ozone from the stratosphere, including methyl bromide, carbon tetrachloride, 1,1,1-trichloroethylene, chlorine, bromine and fluorine. The greatest damage is thought to be caused by the CFCs, used as refrigerants, aerosol propellants and foaming agents in the manufacture of expanded plastics. The two most important commercial products of CFCs are:

R11, which has the chemical formula $CFCl_3$; and
R12 , which is CF_2Cl_2.

It is known that the CFCs and nitrous oxide interfere with the ozone equilibrium in the stratosphere and, although the concentration of these gases is small, it has been shown that the concentration is increasing in line with known emissions of these materials. The precise effect on the ozone layer is disputed, but a number of worrying factors have emerged in recent years, perhaps the most significant of which is the hole in the ozone layer over the Antarctic. It has been estimated that, over Europe and North America, the concentration of ozone is decreasing by 2 to 3 per cent per decade (Harrison, 1992, pp. 13–15).

The possible effects of ozone depletion include increased levels of skin cancer and other skin infections, a contribution to global warming, lower crop yields and damage to fish stocks.

Acid rain

When fossil fuels are burned, in addition to the release of carbon dioxide (CO_2), a number of other oxides are also released, including sulphur dioxide (SO_2) and the oxides of nitrogen, nitric oxide (NO) and nitrous oxide (NO_2), often referred to collectively as NO_x. These oxides are discharged into the lower atmosphere, where they may undergo chemical reactions which convert them into sulphuric acid (H_2SO_4) and nitric acid (HNO_3). These acids cause a decrease in the pH (i.e. an increase in acidity) of water held in the atmosphere. When this moisture is precipitated, it results in 'acid' rain. The natural pH of rainwater is between 5 and 6. The term 'acid rain' is used to indicate rainwater with a pH of less than 4.5. The policy of increasing the height of chimneys on power stations has been thought to contribute to the dispersion of acid rain to areas many hundreds of miles away from the source of the pollution (Rose, 1991).

It has been suggested that acid rain is responsible for the increase in acidity observed in lakes and rivers, although this link has been disputed. Certainly there has been damage to lakes and streams, to forests and to stonework on buildings.

Freshwaters

Acidification of water

This was discussed above, under the subject of acid rain. There is clear evidence that the acidity of many lakes and streams is increasing and this, together with the resultant loss of aquatic life, is cause for concern. The loss of plants and fish from lakes is also a threat to the bird population.

Nitrates in water

Particularly in areas of the world where agricultural practice has been to use large quantities of nitrogen-based fertilizers on the land, a resultant problem has been the increased level of nitrates in water supplies. This has affected rivers, lakes and underground water supplies.

Industrial and domestic effluents

Rivers have always been used as a means for the disposal of industrial and domestic wastes. Much damage was done to the waterways of the industrial world through the build-up of toxic wastes, the presence of harmful micro-organisms and the deoxygenation of the water, and the resultant loss of fish and other aquatic life. While much has been done to clean up rivers, work is still required. In the UK, the Water Act 1989 charges the National Rivers Authority with responsibility for the control of the quality of river water. Measures of water purity in rivers include: Biological Oxygen Demand (BOD) a measure of the extent to which impurities in the water remove dissolved oxygen, essential for much of the life in the river; suspended solids; heavy metals; harmful bacteria; and nitrates from industry and agricultural fertilizers.

The oceans

As with the rivers, the oceans have always been seen as a vehicle for disposing of waste materials. It was considered that they were so large and the biological processes sufficiently robust, that the dilution effect together with the ecological processes would render waste harmless. It is now realized that dilution cannot be relied on to disperse dangerous materials and that the ecological balance of the oceans can be disturbed by the presence of industrial and domestic waste materials. Other forms of pollution include oil slicks resulting from damaged oil tankers, such as the *Exxon Valdez* running aground in Alaska in 1990 and the presence of Tributyltin (TBT), used as a marine anti-fouling agent.

The land

One of the major effects of industry has been the generation of industrial waste sites, which are often heavily contaminated. Sources of contamination of the land include: materials deposited from the air or as a constituent of rainwater, such as acids and heavy metals; agricultural products (such as fertilizers and insecticides); materials from derelict industrial sites; and waste disposal such as landfill sites. The majority of all waste disposal is by landfill, even though a large proportion of this waste could be recycled.

Another important factor in evaluating environmental impact has been the effect of deforestation and loss of land from agricultural use on the reflectivity of the earth's surface. Land which is covered in a tropical rainforest reflects only 10 per cent of the radiation from the sun, the majority of the radiation being absorbed by the vegetation (Goudie, 1993). Where ground is covered in light woodland, grass or heather 15–20 per cent of the energy is reflected. In urban areas, 17 per cent of the energy is reflected. In contrast, a desert region reflects almost 40 per cent of the energy. The more energy that is reflected, the more that energy is likely to contribute to global warming. Loss of tropical rainforests are also causing concern in terms of the loss of habitat for numerous plant and animal species and the resultant loss of biological diversity. Rainforests also have a stabilizing effect on the climate in various parts of the world. Climate changes are also causing a loss of grazing lands and an increase in the desert areas in sub-Saharan Africa. Overcropping and the use of artificial fertilizers are causing soil erosion in some parts of the world.

A third factor is the depletion of mineral resources which exist in the earth's crust. This includes the fossil fuels, responsible for large proportions of energy supplies together with the raw materials for much of the plastic, chemical and pharmaceutical industries, minerals and building materials.

Urban effects

Where people live together in dense communities there are a number of related environmental effects. For example, most cities have higher day and night-time temperatures than the surrounding rural areas. Cities absorb large amounts of heat during the day and release this heat at night. These temperature changes can result in higher rainfall and a greater likelihood of thunderstorms.

One of the other major impacts of cities is the generation of photochemical smog and smoke haze. Much of the smog which occurred in the UK in the 1950s and 1960s was a result of the combustion of fossil fuels, particularly those containing large amounts of sulphur dioxide, and the Clean Air Act 1956 was very effective in controlling this (although possibly with the result of acid rain from the use of higher chimneys to disperse the harmful emissions).

Photochemical smog is created by complex chemical reactions between the oxides of nitrogen, sulphur dioxide, oxygen, water vapour and sunlight. Although it does not cause the same loss of daylight, as was the case with conventional smogs, it does have a number of harmful effects, including damage to plant leaves and the health of humans, particularly those who are asthmatic. In urban areas photochemical smog results mainly from road transport (Rose, 1991). Vehicles fitted with catalytic converters can break down NO_x, one of the main contributors to photochemical smog. However, the main problem is the ever-increasing level of road traffic. Changes such as lead-free petrol and catalytic converters can make only small improvements.

Changes in the size of populations and the migration of people from rural areas to the city are exacerbating problems of poverty, starvation and disease in some parts of the world.

Other factors

Radioactive wastes have become a serious problem resulting from the increased use of nuclear energy as a means of generating electricity. One of the biggest problems with radioactive waste is that it decays at a very slow rate and needs storing in safe conditions for very long periods.

Pesticide residues result from their use in agriculture, horticulture and gardens. In theory, pesticide residues should not be a problem, but incorrect use allows them to leak into the environment. One of the major issues is the contamination of food supplies with pesticide residues (Rose, 1991). They can also get into water supplies.

There are many other environmental concerns, which relate to Third World issues, such as insufficient food supplies and the lack of sufficient sources of clean water for direct consumption and agriculture. Many of the global effects referred to above are affecting the poorest people in the world and are contributing to Third World poverty. This in turn causes people to take short-term measures, such as overgrazing of land, destroying the rainforest, or overfishing the sea, in order to survive. It is difficult to pay attention to long-term global threats when you are concerned with the survival of yourself and your family.

Tourism, hospitality and the environment

Does the hospitality industry cause major damage to the environment?

Throughout the 1980s and 1990s, environmental pressures have affected a much wider range of industries. Initially the concern was related to those industries which caused direct pollution of the environment. Now the problem is much broader and relates not only to outputs but also to the whole operation (Elkington and Knight, 1992, p. 14).

The hospitality industry is an interesting case in that it exposes many of the conflicts which arise in implementing environmental policies (Kirk, 1994). For example, many hotels and restaurants are situated in areas of outstanding natural beauty, in historic cities and in regions with a delicate ecological balance. Will the addition of new hospitality facilities attract visitors to areas which already suffer from too much tourism? For this reason there are serious planning considerations when developing a new hospitality facility. Once the facility has been built, what will be the impact of the operation on the local and global environment?

We must also consider customers, many of whom seek as part of the hospitality experience, to be pampered with lashings of hot water, high-pressure showers, freshly laundered linen, an ample supply of towels, copious supplies of food and drink, the availability of swimming pools and saunas and the limousine to take them to the airport. Whatever we do to reduce the environmental impact of hotels can only be either with the consent of customers or in such a way that they do not suffer any perceived hardship.

The location of the hospitality operation, as is the case for most service industries, is fixed by customers' needs and therefore it cannot be situated where there will be minimal effects from traffic, cooking smells and the noise of the disco and other adverse outputs. This local environmental pollution may not be as important as the issues considered by the Rio Conference, but it does affect people's attitudes towards the industry.

The hospitality industry does not cause gross environmental pollution nor does it consume vast amounts of non-renewable resources and therefore it may not be in the front line of environmental concern. It is made up of a large number of small operations, each of which consumes relatively small amounts of energy, water, food, paper and other resources, and which add only a small amount of pollution to the environment in terms of smoke, smell, noise and chemical pollutants. The industry employs 10 per cent of the population and can have a major impact in developing awareness and good practice.

However, if the impact of all these small individual operations is added together, the industry does have a significant effect on global resources. This is the dilemma. How can we persuade companies involved in the hospitality industry (many of them small, independent operators) to take environmental management seriously? Between the push of legislation and the pull of consumer pressure groups, compounded by the cost savings which can result from reducing waste, many companies are now taking environmental management seriously (Goodno, 1994).

What damage do hotels do to the environment?

CO_2 emissions
CFC emissions
Noise, smoke, smells
Health of staff
Waste energy
Waste water
Waste food
Waste disposal
Agricultural ecology (local produce reduces energy and supports local industry)
Purchasing polices
Transportation policies
Sale of souvenirs made from endangered species
Location of hotels in fragile locations.

In the 1993 Annual Report of the World Travel and Tourism Environmental Review (WTTERC, 1993) there is a recognition that environmental issues will become much more prominent as a factor which influences consumers, regulators, pressure groups and destinations and that the tourism industry will need to show increasing concern for these issues. The report suggests that this can be done by:

● Developing clear policy and mission statements on the environment
● Establishing targets which can be assessed, covering waste management, energy, emissions, hazardous materials, water, noise, purchase of materials and transport
● Disseminating environmental awareness throughout the company
● Encouraging education and research into improving environmental programmes
● Putting an emphasis on self-regulation but recognizing the need for national and international regulation.

Hotel developers have a responsibility to create hotels which are sensitive to the local culture and architecture. What is suitable for an artificial environment such as Disney World may be entirely inappropriate in a historic city centre. The city-centre hotel should reflect local factors but, in practice:

> High rise hotels are indistinguishable one from another, from Tokyo to the Spanish beaches. Tourism has changed the lives of cities; from active participants, the buildings occupying the great public spaces have been turned into passive spectators, the landmarks stripped of any other significance, but that of draws for the coach parties (Sudjic,1992, p. 263).

Hotels, in general, have been slow to develop an interest in environmental management other than through cost-motivated energy management. However, the launch of the Hotels Environment Initiative by the Prince of Wales on 31 May 1993, in which eleven of the top international hotel companies worked together (IHEI, 1993), acted as a catalyst. Since then, the Hotel Catering and Institutional Management Association (HCIMA) and the World Travel and Tourism Council (WTTC) have agreed to co-operate in developing acceptance of environmental issues throughout the hotel industry (Anon., 1994). The WTTC have developed a strategy, known as the Green Globe, in order to promote environmental management among hotel and travel companies (Selwitz, 1994). A number of hotels have started initiatives to protect the environment (Hasek, 1993; Goodno, 1993; Rowe, 1992; Iwanowski and Rushmore, 1994; and Dempsey, 1993). These vary from waste and energy management to the development of eco-hotels and the classification of hotels on the basis of environmental impact.

What are the incentives for hotels?

Some environmental developments which result in reduced consumption of resources such as energy, water and food may also save money (Iwanowski and Rushmore, 1994). However, according to Woodward (1994), these financial savings which may be achieved by developing environmental policies and procedures should not be seen as the main incentive.

Some environmental policies may increase costs which then must be passed on to the customers. Would this act as a disincentive or is it possible to use a 'green' image as a way of marketing a hotel (Feiertag, 1994)? A number of US hotel groups have generated consumer interest by publicizing the fact that they have environmental policies. In a survey of US business travellers (Watkins,1994) 75 per cent of the sample said they were environmentally minded consumers and 54 per cent said they were environmentally minded travellers. Of the sample, 71 per cent said that they would prefer to stay in hotels that show concern for environment. However, a majority were not willing to pay extra for their accommodation in order to fund these green policies and only 28 per cent would be prepared to pay between $5 and $10 extra.

References and further reading

Anon, (1994). HCIMA backs new environmental awareness initiative. *Hospitality*, August, 16–17.

Blowers, A. (ed.) (1993). *Planning for a Sustainable Environment*, London: Earthscan.

Dempsey, M. (1993). Hoteliers convert trash into treasure. *Hotel and Motel Management*, **208**, part 15, 27 and 33.

Department of the Environment (1992). *The UK Environment*, London: HMSO.

Elkington, J. and Knight, P. (1992). *The Green Business Guide*, London: Victor Gollancz.

Feiertag, H. (1994). Boost sales with environment-driven strategy. *Hotel and Motel Management*, **209**, 2, 8.

Goodno, J. B. (1993). Leaves rate Thai hotels on ecology. *Hotel and Motel Management*, **208**, part 7, 8 and 52.

Goodno, J. B. (1994). Eco-conference urges more care. *Hotel and Motel Management*, **209**, part 1, 3 and 22.

Goudie, A. (1993). *The Human Impact on the Natural Environment*, 4th edition, Chapter 7, Oxford: Blackwell, pp. 302–348.

Harrison, R. M. (1992). *Understanding our Environment: an Introduction to Environmental Chemistry and Pollution*, 2nd edition, Cambridge: Royal Society of Chemistry.

Hasek, G. (1993). Waste-removal remedies. *Hotel and Motel Management*, **208**, part 19, 89–90.

HMSO (1990). *This Common Inheritance*, White Paper, London: HMSO.

HMSO (1994). *Sustainable Development: the UK Strategy*, London: HMSO.

IHEI (1993). *Environmental Management for Hotels*, Oxford: Butterworth-Heinemann.

Iwanowski, K. and Rushmore, C. (1994). Introducing the eco-friendly hotel. *Cornell Hotel & Restaurant Administration Quarterly*, **35**, 1, 34–38.

Kirk, D. (1994). Environmental Management – the case of the hospitality industry. Professorial lecture, Queen Margaret College, Edinburgh, 13 December.

North, R. D. (1995). *Life on a Modern Planet*, Manchester: Manchester University Press.

Rose, C. (1991). *The Dirty Man of Europe*, Chapters 4, 6 and 10, London: Simon & Schuster.

Rowe, M. (1992). Greening for dollars. *Lodging Hospitality*, **48**, part 12, 76–78.

Selwitz, R. (1994) WTTC tax fight takes 'comic' turn. *Hotel and Motel Management*, **209**, part 7, 1 and 44.

Sudjic, D. (1992). *The 100 Mile City*, Chapter 10, London: André Deutsch.

Watkins, E. (1994). Do guests want green hotels? *Lodging Hospitality*, **50**, 4 70–72.

Welford, R. and Gouldson, A. (1993). *Environmental Management and Business Strategy*, London: Pitman.

Woodward, D. (1994). Is going green cost effective? *Voice*, September, 18.

World Commission on Environment and Development (1987). *Our Common Future*, Oxford: Oxford University Press.

Wright, D. (1992). *Philip's Environment Atlas*, London: Philip.

WTTERC (1993). *World Travel and Tourism Environmental Review – 1993*, Oxford: World Travel & Tourism Research Centre, Oxford Brookes University.

Young, S. C. (1993). *The Politics of the Environment*, Manchester: Baseline Books.

2 Environmental management

The environmental system

Environmental management is a broad term, covering issues such as:

- Environmental impact – aesthetic, cultural, ecological, social and political
- Sustainability
- Resource and waste management
- Control of emissions and pollution.

Environmental issues are often complex, involving a large number of interactions so that cause and effect are often hard to visualize. In order to understand this complexity, it is useful to consider the organization as a system. While a reductionist scientific approach helps us to understand the detailed working of all various parts, it ignores the complexity of interactions between these parts. We have to establish the relationship between the global effects and how our action at a local level impacts on these global issues. As Hawken (1993, p. 201) states: 'Most global problems cannot be solved globally because they are global symptoms of local problems with roots in reductionist thinking that goes back to the scientific revolution and the beginnings of industrialism.'

At a very simple level we can consider a system as having three essential components: inputs into the system, outputs from the system and the system itself, in which transformations are made in converting inputs into outputs (see Figure 2.1). When we apply this to a hotel we may consider the components as:

- Intangible inputs – clients who come to obtain a specific service; employees who are looking for a rewarding job
- Physical inputs – food, raw materials, consumables, energy, water, etc.
- Financial inputs – the providers of capital who are looking for a return on investment
- The system: physical
                 human
                 technical
                 financial

- Outputs (useful):   satisfied guests
　　　　　　　　　　　satisfied employees
　　　　　　　　　　　sound financial returns;
- Outputs　　　　　　waste materials
  (undesirable):　　　waste energy
　　　　　　　　　　　environmental pollution.

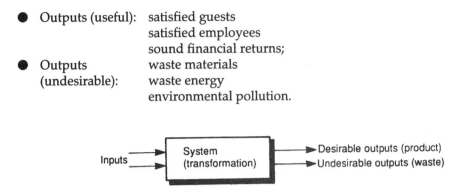

Figure 2.1   *Schematic diagram of a system*

In the study of systems we can think of the system being separated from its environment by a boundary. Here the meaning of the term 'environment' simply means anything which exists outside the system. In looking at the relationship of components within and between systems we can think of two different types of system: closed and open. As the name implies, closed systems have boundaries which prevent any interaction with their environment; they are totally impermeable (see Figure 1.2, Chapter 1). Open systems, in contrast, have a porous boundary which facilitates interaction with the environment. This interaction is essential in the case of any human activity system such as a hotel, but we need to be able to control the flow through the boundary to allow beneficial impacts on the environment and to minimize harmful impacts.

If we contrast natural ecological processes with business we can see some significant differences (Hawken, 1993, p. 12). In an ecological system there are equilibria established such that the waste from one process becomes the input or raw material of another so that chemicals like nitrogen, carbon, oxygen and water are constantly recycled, with little loss or waste from the system. In contrast, industry produces waste which at best is a loss to the system and, at worst, can be toxic and damaging to it. If we look at energy, natural systems use direct energy from the sun (a renewable source) as the primary source of energy which is cycled through plant and animal systems. Businesses on the other hand, consume stored forms of energy (non-renewable) which, once they are used up, will not be replaced.

In the case of any resource, be it food, energy or water, there is a natural equilibrium which determines the rate at which that resource is being replaced. If an activity of commerce such as farming or fishing removes the resource at a faster rate than that at which it can be replaced, stocks of that resource will be diminished and the process will be non-sustainable. A useful concept when looking at the natural balance is that of 'carrying capacity'. This relates to the level of activity which a system can withstand without there being a loss of equilibrium. In a city which is a major tourist destination the carrying capacity might be related to the availability of transport, hotel beds or food supplies. If tourism develops beyond the carrying capacity of that city there will be a deterioration in the performance of the system and a loss of resource or amenity. Other limits might include water supplies, atmospheric emissions, effluent and waste disposal. If the system (in this case the city) cannot cope, it may increase its carrying capacity by importing food and water from outside the system or export its wastes far away. This may have the effect of increasing costs and spreading the environmental impact.

The above is a simplistic analysis since, in the real world, systems are complex and are made of many separate components or sub-systems and these sub-systems will have interactions which have an effect on the environment. As an example people who control the financial sub-system may put pressure on the purchasing sub-system which may encourage them to purchase foods in large quantities because of the supplier's discount structure. This may lead to some food being stored for long periods of time which may increase food waste in the kitchen sub-system. In this case, although the symptom of the problem appears in one sub-system (the catering system), its cause lies within a different one (the financial system).

Clearly, there is a need to manage the inputs and outputs of the system in order to maximize positive impacts and minimize negative impacts on the environment. In global terms, there is disagreement between those experts (Meadows *et al.*, 1972) who predict through an extrapolation of current rates of consumption of raw materials and resources that current growth rates are unsustainable and those other experts who feel that people have always been able to find alternatives when a particular resource becomes limited or its carrying capacity is exceeded (North, 1995).

The organization has a large number of responsibilities in terms of the protection of the environment. It should have sound purchasing policies and choose suppliers who have a similar commitment to the protection of the environment. The environmentally aware organization will control the types of interactions which take place within the system but it must also be aware of any impact on the environment.

An ecosystem is defined in terms of the complex interactions between the flora and fauna in an area, living together in a shared environment in such a way as to produce a balance or equilibrium. An ecosystem is self-regulating in the sense that energy sources and other resources are self-sustaining and wastes produced are reprocessed within the system. The term has been extended to include human activities as a part of natural ecosystems.

Location of hotels in the countryside, areas of outstanding natural beauty and in wilderness areas can damage the very things which make them attractive to tourists. The physical design should be such as to minimize the use of non-renewable materials in its structure and to minimize any negative effects on the environment, minimize waste in its operations, protect employees and guests from any hazards and minimize any harmful or unpleasant emissions. Many companies have sound policies, but all must comply with minimum standards as defined by external control agencies which come through official channels such as legislation as well as via consumer pressure groups.

Environmental policy, strategy and implementation

Fitting into the organizational culture

In order to understand how we make changes in an organization we need to understand something of the way in which it works. Simply by establishing a policy on environmental management we cannot be certain that this will result in any meaningful change. In fact it is highly unlikely that a policy by itself will have any long-term effect.

Hotels consist of many different groups of people including not only the staff who work in them, but also the customers, the suppliers, the local community, people who finance the hotel, and local and national government who establish policy and who are also responsible for many environmental services. It is recognized that in order to introduce change there has to be an acceptance of the reasons behind the change from all involved (Leslee, 1993). Of prime importance is the support which must come from the very top of the organization. It is generally recognized by writers such as Elkington and Knight (1992, p. 34) that, to be totally effective, the adoption of environmental policies must be supported throughout the organization. Without a commitment at the highest level of the company, it is unlikely that ideas developed throughout the organization will flourish.

Once it has been possible to define policies related to environmental management, this must be followed by the establishment of agreed objectives and the provision of any necessary resources which are required to achieve them. These objectives should lead to an agreed programme of changes which should be communicated to all those involved. Conflict can often arise at the level of middle management, where there may be a disagreement between compliance with environmental policy and that of other policies of the company, particularly those involving cost or operational targets. It is at this time that support from senior management becomes critical.

In order to implement a programme of environmental management, it is necessary to identify a group of staff who will plan this implementation and who will communicate both the reasoning behind the plan and its details to all staff involved and who will act as champions of the plan. This group will need to fit in with the relevant structure and culture of the organization. They may be considered as:

- A committee
- A working group or
- A problem-solving group.

The group should be convened by a senior member of staff. It is often helpful if this person has responsibility for the implementation of environmental policy as part of their job specification. The success or failure will, to a large extent, depend on the leadership provided by this convenor and also by the support given by the general manager to the group.

The group, which should be small but drawn from different departments, can be supplemented by other members of the hotel on an *ad hoc* basis as and when required. It cannot work in isolation and must communicate well with all of the organization. A group which is working effectively will motivate individuals to make changes and communicate the value of these changes to the rest of the workforce. To this end, the functions of this group have been defined as being to:

Raise awareness
Build commitment
Provide support
Reward and recognize efforts
Celebrate success

Another important aspect of policy is to develop partnerships with suppliers, guests and the local community. By acting with all these groups, the hotel will maximize the benefits of any action taken, explain the reasons for changes to guests and win the support and commitment of the community.

Environmental policy

A clear statement of environmental policy is the first step any organization that wishes to introduce environmental management should take. The presence of a policy shows a clear commitment to the attainment of targets against which performance can be judged and the organization and its employees made accountable. Because of this, any initial move towards an environmental policy must be realistic in terms of targets since an overblown statement will quickly be discredited and, in all probability, abandoned. On the other hand, it must be written without ambiguity and with sufficient detail to make responsibility for action clear.

The first stage in developing a policy is to identify any laws or codes of practice which can be considered as representing minimum standards. Similarly, any company policies on environmental management need to be incorporated. Major customers or clients may have their own policies which may influence policy, as may any statements from relevant trade and professional bodies.

The policy should cover all aspects of the operation, including marketing and purchasing as well as the more obvious areas of operational management. While a policy statement will vary from one company to another, it should cover a requirement for an environmental audit and the establishment of key performance targets that are regularly monitored and against which any programmes can be evaluated. It should incorporate the development of partnerships with trading partners, including contractors and suppliers.

Some companies have already incorporated environmental values in their mission statements. To be effective, this vision must be converted into clear objectives and targets together with effective monitoring, control and communication. This is equally true for the hospitality industry. The HCIMA Technical Advisory Group issued a Technical Brief on Environmental Issues. According to this brief:

Every business should have a policy statement which should as a minimum make a commitment to:

> The concept of sustainable development;
> Practical action to protect the environment.

In order to protect the environment a business must balance two often-conflicting requirements, that of satisfying the needs of existing customers while at the same time protecting the needs of future generations. Sometimes this can be done through efforts to improve efficiency, thereby reducing the waste of scarce resources. These changes may result from market forces, but often it requires other forms of pressure such as national policy and legislation or public opinion.

Individual companies are linked to a number of other organizations and individuals and an important aspect of environmental management is to form strategic alliances.

Organizations involved would include trading partners such as suppliers, distributors, advertisers, public relations companies and retailers of the service or product, together with groups such as professional and trade bodies. Individuals would include investors, employees, customers and those individuals who make up the community within which the business operates.

The principles of sustainable development have become a cornerstone of moves towards environmental management. These principles have been expressed in a number of different ways and cover the maintenance of physical, cultural and sociological resources, so that they can be passed on to future generations. In terms of physical resources, sustainability may be related to controlling the rate at which non-renewable resources are destroyed. In this context materials like fossil fuels and minerals constitute a finite resource: only a fixed amount is available within the mineral resources of the earth. The more of these materials used by current generations, the less is available to future generations.

In order to embrace the principles of sustainable development businesses must provide goods and services which satisfy the needs of their customers with the minimum environmental impact during all stages of business.

Companies may respond at many levels to the challenges of environmental management. At a strategic level, they may seek a competitive advantage through aspects of product design as well as through the development of market opportunities and the development of staff loyalty and enhanced community relations. At operational management levels, companies may seek to incorporate energy management as a means of reducing waste and improving product quality.

Legislation

Initial policies from organizations such as the European Community were related to major pollution caused by heavy industry and areas of common concern such as controls on the use of lead in petrol. By the 1990s, the issues had become much broader and were concerned equally with the development of principles and policies as well as on developing action against specific perceived problems. The importance of the consumer was also recognized by introducing the concept of ecological labelling. Many of these ideas are enshrined in the Fifth Environmental Action Programme of the European Community. While this policy document is concerned with the development of a framework of legislation based on EC Directives, it also recognizes the importance of commercial pressures on environmental management through imposing realistic financial charges on waste disposal and effluent discharges into waterways. This approach incorporates the principle that 'the polluter pays', through measures such as a 'carbon tax' on fossil fuel sources.

Within the UK, the Environmental Protection Act 1990 brought together much of the existing legislation on the environment together with implementation of key EC Directives. While it is true that much of this legislation is only of concern to those heavy industries which cause gross pollution of the environment, it does impose on local authorities responsibilities for controlling activities such as the regulation of emissions to the air. The control of liquid emissions is charged to the National Rivers Authority. As part of its commitment the UK has developed a White Paper on sustainable development and an agenda for action (HMSO, 1994).

Environmental impact assessment

Environmental impact assessment (EIA) is designed to be proactive and preventative in nature. Environmental assessment normally refers to procedures to be undertaken before launching into a new venture but it can also be used to assess the impact of existing operations. It requires the collection and analysis of information about any possible environmental effects. The methodology has developed from planning procedures, where these environmental impacts include the physical, social, political and economic environment of the area affected by the project (Blowers, 1993). However, in the context of environmental management, the analysis needs to be broadened in order to encompass impacts on global issues.

For large industrial and infra-structural developments, such as power stations, airports or motorways, an EIA is a legal requirement, but it is good practice to carry out an EIA for all developments. An EIA requires the proposed design to be measured against the local and global environment in order to assess any impacts that the development might have. Impacts will include positive factors, such as a contribution to the local economy, increased employment opportunities, amenity development, the development of local tourism and increased markets for fresh, local food products. However, there will also be negative impacts such as loss of land, increased emissions and effluents, higher noise levels, increased traffic volumes and the consumption of resources.

The existing environment can be characterized in terms of population demographics, natural flora and fauna, aspects of the landscape such as surface water, water tables, soil types, air quality, climate, socio-cultural factors and existing built environment. Plans can then be developed to minimize the effect of negative impacts while enhancing positive aspects of the development. It is important to differentiate between different forms of impact. For example, there may be a temporary impact during the development phase which is reduced or eliminated once the development is complete. There will be different time-scales of impact covering the short, medium and long term. The impact will depend upon both the nature of the development and the robustness of the local environment. Also there may be indirect effects. For example, it may be proposed that a hotel development in a remote region should promote local crafts through a gift shop as a part of the development. This may have the effect of changing the nature of local crafts from the production of artefacts with cultural or religious significance into the production of cheap facsimile goods.

Once the impacts have been identified, consideration should be given to ways in which they may be minimized through product design and technology. For example, a hotel development in the cultural or historic centre of a community should be designed so that its architecture is sympathetic to local style and materials. In areas where water supply is a problem, water control methods such as recycling can be built into the design in order to minimize impact.

Although most hotel-related projects will not formally require an EIA (unless they are part of a much larger development such as an airport), there are several advantages at an operational level from carrying out an EIA which make the activity worth-while. Initially, it causes management to be proactive and to identify and address possible causes of complaint from the local community and clients rather than waiting for these to happen and then attempting a resolution at this point. Many aspects of environmental management are cheaper and more convenient to install at the planning stage than they are to address retrospectively.

The adoption of an EIA indicates a firm commitment to environmental management from board level and this can be a powerful signal to all members of the organization. Against these benefits must be balanced the cost of introducing an EIA into the planning cycle and the time it will take, which may cause delays in the project.

For an existing operation, the EIA might better be thought of as an Initial Environmental Audit (IEA). Before commencing on a programme of action a company needs to establish baseline standards of performance against which the outcome of these actions may be judged. An IEA will assess the company's current environmental performance and will help to identify short-, medium- and long-term priorities. The initial environmental audit is best carried out by a multi-disciplinary team which includes operational managers, scientific or technical experts and legal experts. It often helps to have an independent environmental consultant as a member of this team. The audit will require extensive data collection, some of which will be based on routine records, such as purchasing records, but other data will require direct measurement and laboratory analysis. The audit requires thorough planning if it is to avoid the dual pitfalls of either being too broad (and therefore excessively time consuming and expensive) or too narrow and missing out key aspects of the business form the data collection. It should cover:

● Physical sites and buildings
● Raw material usage from purchasing to waste disposal
● Energy policy from sources and tariffs through to efficiency, recycling and wastage
● Products, processes and services
● Waste assessment (including packaging, foods, other materials, water) and their recycling or disposal
● Transport systems for internal use and for clients
● Health, safety and accident policy in relation to environmental factors.

This IEA (see Figure 2.2) is different from the periodic audit which forms a standard feature of the environmental programme in that it is an initial one-off review of the 'starting-point' performance. The periodic audit, which follows on from the IEA, has performance targets against which the programme can be judged and which can be used to monitor progress from one period to the next. In the initial audit the only targets available will be standard industrial norms or, in the case of larger groups, performance of similar operations. The initial audit is a key stage in developing both policy and the action plan for the first year. In later years, the periodic audit is a means of monitoring performance against specific targets, reviewing the action plan and establishing new programmes and performance targets.

Targeting and monitoring

Following on from the environmental audit, a series of targets should be developed, set against the statement of policy for the organization. After the initial audit, these targets will be fairly broad but after the first periodic audit and review, short-, medium- and long-term targets can be developed. Where possible, quantitative measures should be established as targets. The achievements of the organization against these stated targets can be monitored on a regular basis as part of the periodic audit. A number of teams will be needed to progress different targets contained within the action plan (see Figure 2.3).

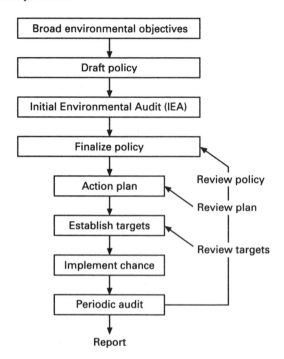

Figure 2.2   *The Initial Environmental Audit and periodic audit*

Short-term priorities will be a mixture of those actions which are essential because otherwise they would impose serious threats on the company (such as compliance with national standards, local codes and conditions of planning permission) and other

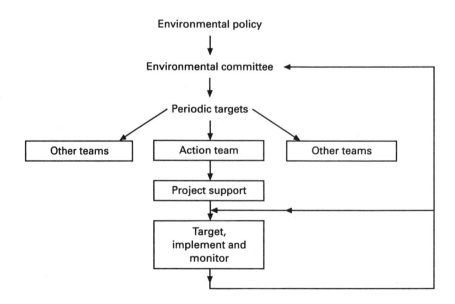

Figure 2.3   *Action teams*

actions which will have significant impact but which are relatively easy to achieve. In the early years of an environmental management programme it is important to build confidence through producing changes which have a high visibility, a short payback period and a low risk of failure. Later, once confidence has been established, more difficult and long-term projects can be tackled.

Key performance measures

The periodic audit should include measures of all key performance measures. While each company may produce a different set of targets, there are a few basic principles when choosing these measures. First, it should be possible to measure these factors (which are related to environmental objectives) in an accurate, reliable and reproducible manner, using either qualitative or quantitative techniques. Second, the number of measures chosen should be sufficient to cover all significant activities of the business but not so many that there is high cost and/or confusion when interpreting the results.

Environmental management systems

The complexity of the environmental management procedures within the organization will depend upon a number of factors such as the stage of development of the environmental plan, the location of its hotels and the size of the company. Systematic procedures, such as those in BS 7750, are available for large organizations and may be suitable for those companies using BS 5750/ISO 9000 for quality management since there are many parallels between the two standards. However, these procedures are very sophisticated and require a significant volume of documentation and record keeping. This may be too expensive to contemplate for the small company, particularly those new to environmental management. However, large chains may find these written procedures essential in order to obtain consistent commitment and performance across the organization.

Case studies

Introducing an environmental programme

Accor (UK) Management Limited is in the initial stages of drawing up environmental programmes tailored to each of its hotel brands. The Novotel UK brand's ecology committee has set out preliminary guidelines under the advice of an ecology consultant.

The programme's long-term goal is to 'ensure that an environmental conscience pervades all aspects of our business – product, people and practice'. The key concepts are:

- Rethinking
- Re-using
- Reducing

- Rationalizing
- Recycling
- Recovering

Novotel selected one person to be the ecology representative for the UK. This person coordinates the programmes in the UK, and is responsible for consolidating quarterly ecology reports received from each hotel and judging them to select a winner. The UK coordinator attends meetings with European counterparts and updates the Novotel UK network.

Each hotel was asked to choose a person who would be the ecology rep for that hotel. Novotel was keen that these persons would not be head of department level, but would be self-nominated and self-motivated with regard to the environment. Their responsibilities include producing a three-monthly report on actions taken/actions planned/savings made; liaising with the chef, maintenance technician and housekeeper; attending meetings with the UK representative; updating heads of department and employees; encouraging ownership of ecology in the programme.

To get the whole thing going, the UK representative held a briefing meeting with each hotel rep and, with the help of an external consultant, took them through the aims and objectives of the programme. To help them with the job of transferring the knowledge and enthusiasm to other staff members, each hotel nominee received a set of slides to enable them to brief others. Each hotel rep was also encouraged to make contact with the environmental officer from the local council to get help and advice on recycling, etc.

Novotel has set short-, medium- and long-term goals for the implementation of the programme:

1 *Short-term goals* – identifying and separating recyclable or re-usable materials from other waste, e.g:
   (a) Separated coloured glass
   (b) Breaking down cardboard and putting it in separate bins where appropriate
   (c) Re-using the reverse side of computer/photocopy paper as scrap
   (d) Obtaining estimates for printed material to be printed on recycled paper
   (e) Keeping vegetable waste.

2 *Medium-term goals* – identifying parties who may be interested in recycling/re-using material and discussing possible contracts with them, e.g:
   (a) Setting up a desk collection sytem for waste paper
   (b) Identifying paper merchants/glass companies that would take paper/glass for recyling
   (c) Contacting schools that could use egg-boxes, plastic containers, glass jars, etc.
   (d) Finding charities/recycling centres that take aluminium cans
   (e) Contacting zoos/farmers about taking food scraps
   (f) Finding retirement homes/hospitals to take old magazines or newspapers.

3 *Long-term goals* – identifying areas where arrangements can be made with suppliers or investments can be made in machinery, e.g.:
   (a) Seeing whether glass-reducing machines could be useful
   (b) Investing in composting bins
   (c) Discussing with suppliers the re-use of plastic, glass and cardboard packaging.

Ramada Group

At Ramada hotels around the world, employee-run environmental committees have been set up to develop programmes and generate staff support. The group has also organized an annual Chairman's Award to honour region, hotel and employee for outstanding environmental efforts with cash incentives. Some Ramada employee-based initiatives are as follows:

1 At the Renaissance hotel in Atlanta employees adopt a street and spend 1 hour a week collecting trash.
2 At the Ramada Renaissance in Bloor, Toronto, employees have planted a total of 1500 trees in local parks and adopted an endangered animal at the Metro Zoo.
3 Ramada International employees in the North America Regional Office sponsored a 'Green Fling' environmental education week and donated $500 to an endangered species programme at the Phoenix Zoo and to the Nature Conservancy. Similarly, another group adopted a section of highway for clean-up twice a year and set up a ride-share programme for employees.
4 At some hotels environmental committees are publishing newsletters to educate and inform staff and guests about their efforts.
5 The 'Green Team' at the Ramada Renaissance Resort in Aruba organized a clean-up in the playground of St Dominicus College and sponsored the first Ramada Bicycle Rally to raise AFLS $470 for Green Team efforts. The Aruba group also installed trash bins and developed the 'plant a tree' programme in local schools, and set up a guest awareness programme to protect the environment.
6 The Ramada hotel group distributes to guests leaflets entitled *Fifty Simple Things You Can Do To Save The Earth*.
7 For younger guests, Renaissance hotels in the USA feature a children's menu that encourages children to help protect endangered animal species.
8 Ramada International Hotels and Resorts has also developed a marketing initiative in conjunction with American Express to raise funds for the Nature Conservancy and to inform a wider audience of its work. During the programme, each time accommodation at a Ramada hotel is paid for by an American Express card, the two companies jointly donate $1 to the Nature Conservancy.

The Nature Conservancy, founded in 1951, is a non-profit, global conservation organization. Since its foundation it has helped to protect more than 5.5 million acres of land and save thousands of species, including some 990 on the brink of extinction. The funds raised by the Ramada/American Express initiative are allocated for the preservation of the Palau/Yap archipelago in Micronesia, which is considered to be one of the marine wonders of the world. This archipelago currently supports over 1350 reef fish species, and 80 per cent of the bird species that live on the islands are found nowhere else in the world. The introductory phase of the programme in 1990–1991 raised US$83 000 for the Nature Conservancy, while during the period 1992–1992 this figure reached US$102 000.

Miami Dadeland Marriott Hotel

The 303-room Miami Dadeland Marriott has been running a comprehensive environmental programme since October 1991; it grew out of an initial suggestion from a staff member that the hotel begin recycling. The objective of the programme is to promote environmental responsibility as well as good industry and guest relations. The programme now encompasses the 'Green Rooms' campaign, energy conservation, and environmental fragrancing, which is the use of selected, aromatic oils, known to produce effects varying between relaxing, soothing, invigorating etc. The oils are vaporized and pumped into the lobby and lounge of the hotel.

In response to heightened concern about green issues, the 'Green Rooms' campaign offers guests alternative accommodation. Working with an environmental design group, the hotel has installed low-flow shower heads, faucet aerators and toilet dams to all rooms in order to conserve water. In addition, air and water filtration units have been added to nineteen rooms.

The HEPA (High Efficiency Particulate Arrestor) filtration system is designed to remove most odours, dust, pollen, mildew, dirt, smoke as well as other particulates from the air. Similarly, the water system removes chemicals and disinfectants, providing guests, in effect, with bottle quality water at the tap. These systems are of particular benefit to guests who suffer from allergies, while non-sufferers have also reported noticeable effects.

The Miami Dadeland Marriott publicized its 'Green Rooms' through advertisements in newspapers and magazines as well as using marketing materials within the hotel itself. Although the adapted rooms cost guest an extra $5 per night, the positive response to be campaign has meant that it has now been extended to thirty-eight rooms.

The results of the environmental programme are measured through guest comments, advance 'Green Room' bookings, electricity bills and waste removal records. In order to deal with the increase in environmental enquiries by guests and others, an active task force of hotel employees has been created to train co-workers, answer specific questions and identify new environmental initiatives. The Miami Dadeland Marriott has benefited from local, national and international media attention.

Swiss Hotel Association

The Swiss Hotel Association asks members to encourage guests to re-use bathroom linen. A notice in one of the Accor group's hotels in Switzerland reads:

> Every day, hotels launder a great many towels – most of them unnecesarily.
> This leads to enormous quantities of detergents polluting our water system.
> You too can make a contribution to preserving our environment – by using
> your linen more than once!

Forte plc

With establishments in hundreds of towns across the UK, Forte plc has always endeavoured to take an active part, as a socially responsible company, in conserving and preserving the environment of the communities in which it operates.

Recognizing that good intentions need to be backed by financial support, Forte, in association with the Conservation Foundation, launched a 'Community Chest' Scheme

8 years ago to provide monthly grants to aid local environmental projects of all kinds. In the past 8 years the Community Chest has, among other things, helped to reseed village greens, create school wildlife gardens, plant trees and transform acres of wasteland into community gardens. The number of projects chosen by an independent panel headed by naturalist David Bellamy has now reached over 100, and the scheme's successful century is being celebrated with new literature printed on special recycled paper.

In an effort to encourage greater tidiness on the streets of Britain, the company's Kentucky Fried Chicken has also run for some years the Colonel Sanders Environmental Awards, aimed at encouraging all sectors of the community to take action and help keep Britain tidy. Winners of four categories, local authorities, voluntary goups and two specifically designed for young people, receive a trophy made of wood salvaged from the great storm of 1987.

Amstel Inter-Continental

The Amstel Inter-Continental Hotel in Amsterdam reopened in September 1992 following a US$50 million refurbishment. This success story serves to illustrate that environmental issues are not purely confined to functional operational areas, but architectural and aesthetic considerations are taken into account. The refurbishment programme of the Amstel Inter-Continental was designed to preserve the building's most beautiful features, while modernizing existing facilities. Original brickwork that had decayed was replaced with new work. Casts of existing columns and plasterwork were taken and used to provide mouldings to clad the new columns supporting the roof of the Grand Hall. The 1920s extension for the hotel's bar was redesigned to match the surrounding architecture under the direction of the local authority and the Historical Monuments Commission. Original decor was researched and carefully restored in most rooms. The exterior of the building was carefully repaired and cleaned, using environmentally friendly techniques and meeting standards set by the Historic Monuments Commission.

Environmental considerations are an important part of the day-to-day operations of the hotel. Water from the river is used for cooling compressors – and returned to the river without any additives. State-of-the-art technology ensures that the hotel operates at maximum energy efficiency and waste is minimized. No harmful gases, such as halon (often present in firefighting equipment), are used throughout the building.

Meridien Hotels

Meridien head office felt that telling hotels what to do by sending directives and procedures was neither fair nor sufficient to demonstrate that environmental preservation is everyone's concern. The group also felt that they could not promote environmental responsibility until head office had its own action plan in place. Starting in autumn 1992 they implemented an action plan to gain faith and support from their worldwide staff. The plan mainly relies on 'Reduce; Re-use; Recycle' and energy conservation but will also cover all regular office activities. Starting with an initial basic plan, suggestions from the 100 employees at head office will be encouraged, while a committee will weigh the cost efficiency of each suggestion.

A bulletin *Initiatives Environment* supports and recognizes the best suggestions and efforts. This plan works towards asking regional head offices, sales offices and individual hotels to join the hotel environment preservation plan by mid-1993.

Tamanaco Inter-Continental, Caracas

The Tamanaco Inter-Continental Hotel in Caracas initiated an extensive campaign to raise awareness of its environmental programme, both internally and externally, using stickers, caps, posters and stationery distributed throughout the group's Latin American hotels. As a consequence, an oil company based in Venezuela, itself keen to maintain a positive environmental image, called the hotel to book a conference at which environmental issues were to be discussed. The conference took place over a week during the low season in January 1992, generating 350 room/nights and a total revenue of US$40 000.

Holiday Inn Leicester

Leicester was recently nominated as the environment city of Great Britain. As a mark of this, the Holiday Inn implemented an environmental policy in late 1990, which is not only still in operation but is going from strength to strength.

This initiative included changing to large jars of organic fruit jams, using recycled paper where possible and photocopy paper made from rainforest-friendly wood, installing a new low-energy lighting system and an energy conservator on the boiler. All cleaning materials are biodegradable and available in refillable trigger-spray bottles.

The city of Leicester also benefited as Holiday Inn introduced the first bottle bank and also can-crushing, which were resounding successes. The staff are helped and encouraged in every way, with a 'Green Committee' holding regular meetings, a recycling notice board for stamps, cards, etc. and a Green Quiz for staff. Environmentally friendly products are regularly updated, tested and studied. A letter is left in each room, explaining to all guests what is hoped to be achieved and requesting any remarks. In the three years that the programme has been running, no adverse comments have ever been received.

References and further reading

Blowers, A. (ed.) (1993). *Planning for a Sustainable Environment*, London: Earthscan.
BS 7750 Environmental Management Systems, London: British Standards Institution.
Elkington, J. and Knight, P. (1992). *The Green Business Guide*, London: Victor Gollancz.
Hawken, P. (1993). *The Ecology of Commerce*, London: Weidenfeld and Nicolson.
HCIMA, *Managing Your Business in Harmony with the Environment*, London: HCIMA.
HCIMA Technical Brief No. 13. *Environmental Issues*, London: HCIMA.
HMSO (1994). *Sustainable Development: the UK Strategy*, London: HMSO.
Holliday, J. (1993). Ecosystems and natural resources. In Blowers, A. (ed.), *Planning for a Sustainable Environment*, London: Earthscan.

Leslee, J. (ed.), (1993). Canadian Pacific's green plan. *Hotel and Motel Management*, **208**, part 11, 8 and 48.

Meadows, D. H., Meadows, D. L., Randers J. and Behrens, W. W. (1972). *The Limits to Growth*, London: Pan Books.

North, R. D. (1995). *Life on a Modern Planet*, Manchester: Manchester University Press.

Welford, R. and Gouldson, A. (1993). *Environmental Management and Business Strategy*, London: Pitman.

3 Water management

Water and the environment

All hotels require considerable volumes of water, which is becoming an increasingly scarce and expensive resource. Water supplies vary from one part of the world to another. Within the UK, while there are abundant supplies of water, there are shortages at various times of the year, particularly in the south-east of England, because groundwater supplies become depleted. In other parts of the world, hotels may be located where water supplies are at a premium.

The management of water consumption and water quality is important for a number of reasons:

- Waste water diminishes a scarce resource and costs the hotel money
- Waste hot water wastes not only water but also energy
- Poor quality water supplies can provide a health risk to guests and employees
- Poor quality water supplies can increase the running and maintenance costs of equipment and reduce its life
- Contaminated waste water increases the load on effluent plants and may endanger the water supply of others.

Water is one of the basic requirements for all life on earth. Fresh water is one of the most fragile of the world's resources and is essential for food production and for the health of all living species. A large proportion of the earth's water is in the oceans (97 per cent) and only 2 per cent exists in the form of fresh water and, of this, a large proportion is trapped in the form of ice at the poles of the earth and in glaciers.

Water is cycled from the sea by evaporation from the surface and collects in clouds, from where it is precipitated as rain or snow, some of which falls on land. Here, some of the water is trapped underground in the form of aquifers but much of the water ends up in rivers and lakes and eventually makes its way back into the sea as shown in Figure 3.1. The water is used by animals and plants. Additionally, people have found many applications for water ranging from hygiene, cultivation, transportation, medical applications, industrial processes and sports.

Water which is evaporated is pure, but during the precipitation cycle these pure water supplies become contaminated with a number of undesirable materials. For example minerals are leached from the soil. This has become much more significant due to the use of nitrate-based fertilizers on the land. In Europe, the use of fertilizers has increased by a factor of 5 in the last 50 years. The excess nitrates drain off into lakes,

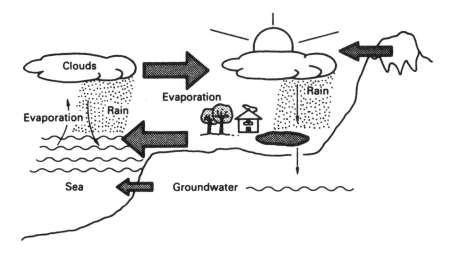

Figure 3.1   *Water circulation cycle*

rivers and underground supplies. With deep water supplies it may take up to ten years for these nitrates to emerge in springs or well-water. These nitrates are known to affect babies and because of this, there are a number of national and international standards for nitrate levels in drinking water (Rose, 1991). Nitrate and phosphate levels have also been implicated in the growth of toxic algae in reservoirs. Sulphur dioxide which gets into the air from the combustion of high-sulphur fossil fuels is absorbed by rainwater and converted into sulphuric acid and results in acid rain. This high acidity enters water supplies via lakes and streams. In addition to the above contaminants in the natural water cycle, rivers are used as a means of waste disposal from farms, factories, homes and sewerage plants. While there are strict controls on discharges into rivers, accidental pollution still occasionally occurs. For these reasons, water usually requires treatment before it is suitable for use for drinking and cooking purposes.

Water supplies

Natural groundwater

Water which falls on the ground in the form of rainfall seeps down through permeable soil until it reaches an impervious layer, to form the bulk of freshwater supplies for all human activity. Here the water collects and fills the space between soil particles or porous rocks, such as sandstone, limestone, sand, gravel and shale. These underground structures are known as aquifers. The depth of this layer of underground water below the surface level of the ground is known as the water table. Water is lost from these underground supplies by leakage through faults in the impervious layer, through springs and from wells and boreholes for human consumption and farming. Aquifers are reached using a shallow or deep well in order to extract the water. One form of aquifer is the artesian well, where the porous rock lies between two impervious layers, often causing sufficient pressure on the water to drive it up to the surface via a well without the need for a pump. The water level in many aquifers is falling, because the rate of extraction is exceeding the rate of replacement (see Figure 1.4, Chapter 1). In some circumstances where supplies are depleted, the fresh water may become contaminated with salt water.

Surface water

Rivers

Rivers are commonly used as a source of water but normally require treatment before use. Particularly in downstream sections, rivers are often contaminated with waste materials from industry, agriculture and communities. In the UK rivers are classified in terms of their quality (Department of Environment, 1992):

- Class 1a, Good Quality: water of high quality suitable for potable supply abstractions; game or other high-class fisheries; high amenity value.
- Class 1b, Good Quality: less high quality than 1a but satisfactory for substantially the same purposes.
- Class 2, Fair Quality: waters suitable for potable supply after advanced treatment; supporting reasonably good coarse fisheries; moderate amenity value.
- Class 3, Poor Quality: waters which are polluted to an extent that fish are absent or only sporadically present; may be used for low-grade industrial abstraction purposes; considerable potential for further use if cleaned up.
- Class 4, Bad Quality: waters which are grossly polluted and likely to cause a nuisance.

Lakes (natural and artificial)

Where there is a shortage of underground water, lakes or artificial reservoirs may be used to provide water supplies but the water usually needs some form of treatment prior to use. Recently, there has been an increase in the occurrence of algal blooms caused by the growth of blue-green algae. Some of these algae produce toxins which are poisonous to fish and mammals. In the UK, the same classification scheme is used for rivers and lakes (see above).

Oceans

The oceans represent the most abundant source of water on the planet, but the cost of desalination is usually prohibitively high and therefore sea water is not often used as a source of water. Coastal waters are often contaminated with sewage and heavy metals.

Potable and non-potable water supplies

The required quality and purity of water depend upon the purpose for which it is to be used. Water to be used for drinking purposes, in ice-making machines, in drinks and for cooking has to be of a very high standard, with controlled levels of harmful minerals and bacteria. It also has to be free of colour, turbidity and odour. Water which is of a standard suitable for drinking purposes is often referred to as being 'potable'. Other 'non-potable' water supplies for cleaning, washing and for use in the garden and grounds do not need to conform to such high standards.

Water supplies in the hotel

Most hotels obtain their water from utility companies. In some remote areas, hotels may draw their supplies from wells bored into aquifers. Where water is scarce, water from roofs and storm drains may be stored in tanks for use on the grounds.

Within the building, water supplies are designed to provide different types of water. A variety of supplies are required (Lawson, 1976) for:

Cold water for drinking
Cold and hot water to bathrooms
Cold water to WCs and bidets
Hot water circulation for space heating
Chilled water circulation for air conditioning
Hot and cold water for kitchen and laundry
Water for firefighting.

In the UK, potable cold water supplies for drinking and cooking purposes will be taken straight from the mains, with no intermediate storage tank. For other purposes

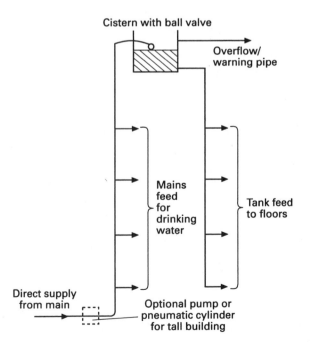

Figure 3.2 *Cold water feed system for mains and tank supplies*

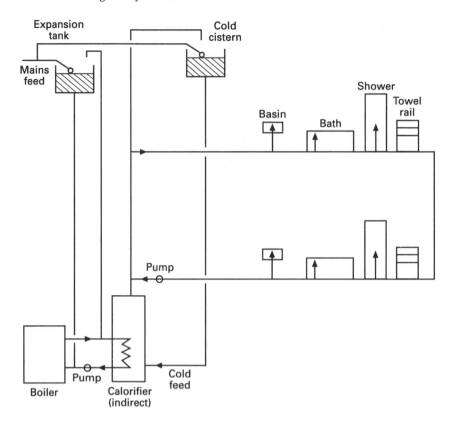

Figure 3.3 *Hot water distribution to bedrooms*

(such as laundry, use in the grounds and swimming pools) cold water supplies will be drawn from a storage tank supply (see Figure 3.2). Hot water supplies are taken from one or more calorifiers (hot water storage tanks). In simple systems, a single calorifier, with thermostatic control, will be used to pipe supplies to all hot water outlets, and a single boiler may provide water for domestic and heating purposes. Water for use in guest rooms will be pumped around a circuit (see Figure 3.3). In larger hotels, duplicated boilers will be used and separate calorifiers will hold water supplies at the relevant temperature for use in guest rooms, cleaning and laundry. Where separate supplies are not possible, localized booster heaters may be used to increase the temperature of the water for specific purposes, such as dish-washing machines. The ideal temperatures of supplies are:

Guest room   50°C
Kitchen (general)   60°C
Sterilizing   80°C
Laundry   80°C.

In addition to the above, there may be other specialized hot and/or cold water supplies for particular appliances, such as softened and pressurized water. The way in which hot water is supplied to each outlet can have a large impact on energy supplies.

The removal of waste water is controlled by local regulations or codes. It is normal practice to separate water which collects from the roof and site drainage (storm drains) from the waste from sanitary fittings (see Chapter 6). In many situations, waste from the kitchens will first pass through grease filters before entering the waste system.

Improving water quality

The nature of impurities in water

Most water supplies must be treated before they are suitable for use in the hotel. This treatment will normally be carried out by the utility company but, in some remote regions, hotels may have their own wells or other local water supplies, in which case some treatment on-site will be necessary.

Water quality can be defined in terms of chemical, bacteriological and organoleptic factors. The desired quality must be related to the actual use of the water supply. If, for reasons of economy and to reduce the volume of chemicals used in water treatment, lower-quality water supplies are to be used for purposes such as in WCs and for use in the grounds and gardens, then it is essential to ensure that there is no possibility of the contamination of potable water supplies with these lower-quality supplies. The two systems must be physically isolated and outlet points with 'non-potable' supplies must be clearly labelled as not suitable for drinking.

In terms of water quality, a number of chemical contaminants of water are of concern. These include lead, aluminium, nitrates and pesticide residues. Historically, much of the lead in domestic water supplies has arisen from the use of lead pipes, although a minority of naturally occurring supplies do contain lead. Lead is dissolved by water at a slow rate, so it is of greatest significance where slow-moving or stagnant water is in contact with lead pipes and tanks. The rate of absorption is accelerated by water which has an acid pH. Lead also gets into the environment through the use of lead-based chemicals as a petrol additive. However, this source of lead is most significant as a contaminant of air.

Aluminium enters water supplies either as a natural component of water which has passed through acid soils or as the chemical aluminium sulphate, used in the clarification of peaty water. It has been suggested that there is a link between aluminium and Alzheimer's disease but this is disputed.

Nitrates in the water supply result mainly from the leaching of agricultural land. They are also present in discharges from sewage plants.

Some water supplies may contain chemicals which results in hardness of the water. This hardness is usually caused by calcium and magnesium salts. Particularly for water supplies to kitchens, laundries, boilers and water-based heating systems, these minerals may need to be removed, a process known as water softening. Hardness also results in scale build-up in water heaters and boilers. Excessive hardness can also cause scum in handbasins and baths when the chemicals in the water react with soaps and, if this occurs, domestic water supplies may also need to be softened.

A localized water softener can be used to remove the calcium and magnesium salts. It consists of an ion exchange resin (zeolite) in the form of small pellets. As water passes over the pellets, sodium ions from the pellets replace the calcium and magnesium ions. After a time, the resin becomes saturated and must be regenerated, by replacing the calcium and magnesium with sodium. This is done by passing salt (sodium chloride) over the resin.

Small amounts of chemicals may cause discoloration or tainting of water supplies, thus reducing their organoleptic appeal in drinking water and ice and in hot and cold beverage making. For example, naturally occurring iron and manganese salts can cause water to have a yellow or brown colour. This may affect the acceptability of drinking water and also the quality of fabrics washed in the laundry. It may also cause staining of sanitary fittings such as handbasins, baths and WCs. Turbidity of water supplies is usually the result of suspended solids in the water. Excessive turbidity may

Table 3.1 *Potable water quality: permissible concentrations as defined by the World Health Organization and the EC (physical–chemical limits and maximum permissible concentrations in parts per million)*

Parameter	WHO standards	EC standards Desirable levels	Maximum levels
Temperature (°C)	10–15		25
pH	6.5–9.2	7–8	6.5–8.5
Conductivity (–S/m^3)		400	400
Chlorides	250	<25	250
Sulphates	200	<250	250
Hardness (as CaCO3)		<100	100
Magnesium		<30	50
Sodium	250	<20	150
Potassium		<10	12
Aluminium		<0.05	0.2
Total dissolved solids (TDS)	1,000	<500	1,500
Nitrates	45	0	50
Nitrites		0	0.1
Ammonium		<0.05	0.5
Aromatic hydrocarbon (fuel)	0.001	0	0.000,2
Phenols	0.001	0	0.000,5
Organic chlorine compounds		0	0.025
Pesticides		0	0.0001
Iron	0.3	<0.05	0.2
Manganese	0.02	<0.05	0.05
Copper	0.05	<0.05	0.1
Zinc	5.0	<0.1	0.1
Lead	0.05	0	0.05
Cadmium	0.01	0	0.005
Chromium		0	0.05
Mercury	0.001	0	0.001
Arsenic	0.01	0	0.05
Cyanides	0.01	0	0.05
Nickel		0	0.05
Silver			00.01

be caused by inadequate water treatment and filtration. Tainting of water can be by natural materials such as peat or because of chemical contamination of the water supply. Very small quantities of some chemicals may lead to an off-odour in the water or in beverages made from the water. Some of these discolorations and taints may be harmless, but all complaints should be thoroughly investigated to ensure that there are no health hazards and that the quality of the water is acceptable to the guest. A specification for potable drinking water is shown in Table 3.1.

Potable water supplies must be free from harmful bacteria. The coliform bacteria are used as indicators of bacteriological water quality. These bacteria, while not normally pathogenic, are associated with pathogenic organisms and are often indicators that a water supply has been contaminated with sewerage. If coliform bacteria are present in water, there is a possibility of faecal contamination and the presence of a number of micro-organisms which cause gastro-intestinal infections. These organisms might include species of *Salmonella* and *Shigella*, *Vibrio cholerae* (which cause cholera), viral Hepatitis A, *Camplyobacter jejuni*, *Camplyobacter coli*, *Yersinia enterocolitica* and some strains of *Escherichia coli*. They are destroyed by heat and chlorine-based disinfectants, but can be a hazard in drinking water, ice and water used in the kitchen for rinsing salads and raw vegetables.

The bacteria *Legionnella pneumophilia* may build up in water tanks, shower heads or air cooling systems and are harmful, causing Legionnaires' disease. It is unlikely to be present in water supplies to the hotel. It can be killed by heat or chlorine-based disinfectants.

The most common methods of raw water treatment are:

- Filtration to remove solids, taste and odour
- Biological oxidation, to remove organic matter including bacteria
- Removal of iron, manganese, acids, odour and taste.

Some substances, such as non-biodegradable organic compounds, heavy metals, phosphates and ammonia, are difficult and extremely expensive to remove.

Chlorination is a common method for the disinfection of water supplies for domestic purposes and swimming pools.

An action plan on water quality

The aim of the action plan is to improve the quality of water by ensuring that no environmental contamination takes place and that any hazard is at least minimized to an acceptable level of risk. This will reduce health hazards to guests, visitors and employees as well as improving comfort. A further aim is to reduce other harmful effects, such as corrosion, scaling and deposits, which will extend the life expectancy of the hotel's equipment and piping systems.

The water quality programme should commence with the identification of standards which are applicable to the hotel and its location. These standards can be international (such as those produced by the World Health Organization and the European Union), national legal requirements and local standards and codes. Where local standards are less strict than those of national and international bodies it is good practice to use the more stringent of the two sets of standards. Supplies of water to the hotel can then be checked against these standards and any remedial action requested from the utility company. If the hotel has its own supply of well water, additional treatment plant may be required to bring the water up to the standard.

The next stage is to check the functioning of existing water plant. Any problems identified may result from poor maintenance or it may be that the plant cannot cope with current standards and may require more serious modification or replacement. Some of the possible quality defects and their treatment are:

- Suspended solids: filtration
- High salt content: desalination
- Iron: potassium permanganate treatment
- Acidity: increase pH
- Corrosivity and scale: chemical treatment
- Hardness: water-softening treatment
- High temperature: cooling
- Bacterial contamination: chlorination and flocculation
- Odour and taste: filtration through an active carbon filter.

All materials and chemicals used in the hotel and which may have a harmful effect on water supplies should be identified. Procedures should be in place to prevent the contamination of water supplies and effluents with these chemicals.

An assessment should be made of the physical plant. Cold water storage tanks are a possible source of contamination, particularly where there is insufficient turnover at all times of the year. The design of tanks should be checked to make sure that the inlet and outlet points are not so close together that stagnant water can build up in the tank. All tanks should be covered and any openings should be designed to prevent contamination from airborne particles, birds and rodents.

Domestic hot water storage tanks should also be checked to make sure that there is sufficient turnover. Water temperatures should be checked to ensure that bacteria, such as *Legionnella pneumophilia*, cannot survive and multiply.

The water distribution system should be inspected to ensure that contamination cannot take place and that harmful bacteria cannot breed in the pipework and contaminate the water. For example, there should be strict isolation of domestic hot water supplies from any other non-domestic systems, which may not operate with the same quality of water as domestic systems. There should be no possibility of back-siphonage from other systems. Where this is a possibility, vacuum-breakers will need to be installed. Any redundant pipework, which will allow the build-up of stagnant water, should be isolated from the distribution system. All kitchen drains should be fitted with grease traps. Any non-potable water supplies for cleaning, irrigation, etc. should be locked (or isolated in some other way) if they are accessible to guests or employees. The removal of waste water supplies is discussed in Chapter 6.

Control of water consumption

Assessing current performance

A water use audit should be conducted which relates measured consumption of water to the time of year and to the level of business in the hotel. Water is used for a wide variety of purposes, as shown in Figure 3.4. Although it provides a detailed evaluation

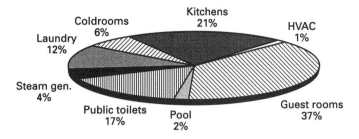

Figure 3.4 *A typical hotel water audit*

of efficiency, the water use audit has little meaning without comparison, either with previous performance figures for the hotel, or by comparison with figures for other hotels. When audits have been carried out for a number of years, this comparison can be of year-on-year performance. Comparison with other hotels can also be an important yardstick (see Table 3.2). To make this comparison, it is necessary to know the annual consumption (C) of water for the hotel and the average number of guests per day (G) for the year. Using this information:

Water consumption (m³ per customer per year) = $\dfrac{C}{G}$

In considering these figures, it is necessary to take account of specific factors which may affect consumption. For example, an in-house laundry may increase water consumption by 25 per cent. Other factors may include the availability of a swimming pool or other leisure facilities and the use of air-conditioning.

Table 3.2 *Typical water consumption (m³ per person per year)*

Type of hotel	Performance			
	Good	*Fair*	*Poor*	*Very Poor*
Large hotel (more than 150 rooms)	<220	230–280	280–320	>320
Medium (50–150 rooms)	<160	160–185	185–220	>220
Small (less than 50 rooms)	<120	120–140	140–160	>160

Based on Tables 3.1 and 3.2 of IHEI (1993).

Changes in occupancy affect water consumption considerably. We might expect that there would be a linear relationship between occupancy and consumption since as the number of occupants increases, more energy and water will be required. However, there is a base load of consumption which will occur if there are no guests in the hotel. Where water consumption is metered it is possible to estimate this base load by collecting data on occupancy levels and water consumption over a period of time, as shown in Figure 3.5. If the base load is high, the reasons for the high level of consumption of water when there are few guests in the hotel should be investigated. Where water supplies are not metered, it may still be possible to estimate consumption through an audit of activities.

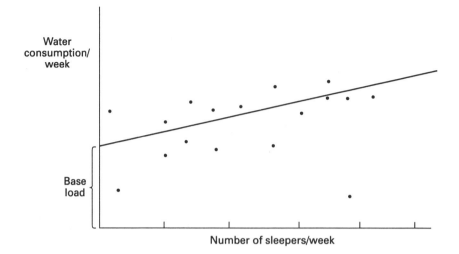

Figure 3.5   *Water consumption and occupancy*

Reducing the wastage of water

Hotels consume large amounts of water and the cost of this can represent a large proportion of all purchases. An additional factor is that there is a close relationship between energy consumption and water consumption. This is because a large proportion of water usage in a hotel is in the form of hot water. If large volumes of hot water are wasted, this will represent a waste both of water and of energy.

In looking at ways of reducing water consumption a key factor is that this must not be done at the expense of the comfort of guests, unless with their overt agreement. Some basic rules for good practice include:

1   Improve efficiency by training all personnel to understand, operate and maintain the hotel's equipment and systems in an energy-efficient manner and to minimize the waste of water.
2   Invest in the building, equipment and systems to improve efficiency.
3   Measure efficiency as a standard procedure, particularly for those areas which are major consumers of water such as boilers, chillers, cooling towers and air-handlers.
4   Set targets for each department and continuously monitor the results.
5   Constantly look for improved technology which is suited to hotel application.
6   Place utility conservation projects on the same level as projects related to interior decoration, structural changes, extensions, additions etc.
7   Good training is the first and most important step and it should be an on-going process for all staff.

At the design stage of a new hotel there are many opportunities for building in the efficient use of water supplies. For example, the desired temperature of hot water varies for different purposes. Hot water for supply to taps in guests' rooms does not have to be as hot as that for many other domestic purposes. Indeed, tap water which is too hot can provide a safety hazard, since there is a risk of scalding. By having separate distribution systems for each type of hot water supply, each with its own thermostatic control, energy consumption can be reduced. Another key factor is the choice between baths and showers. Baths consume considerably more water (and therefore energy) than do showers. Should the design include bath and shower fittings in all rooms, or should some rooms have shower only? This will, to a large extent, depend upon the wishes of guests. Even in rooms with showers, considerable volumes of water and energy can be saved by controlling the water pressure. There has been a trend towards high-pressure showers, provided on the basis of the invigorating massaging effect of high-pressure water jets. Do guests really want this facility or would they settle for a reasonable flow of water, with a resultant saving of water and energy?

At the time of commissioning, all the water systems in the building should be checked to make sure that they are working in accordance with the design specifications. At the time a building is handed over to the hotel company there are many pressures to start business as soon as possible in order to generate revenue. Experience shows that some jobs never get satisfactorily completed or specific problems resolved. If these problems are not addressed at the time of commissioning and hand-over, the hotel will be penalized with higher utility rates for the rest of its life. In the case of hot water supplies, any defects in the distribution system will also result in higher levels of energy consumption.

During the life of the hotel, much can be done in order to conserve and protect water supplies. The achievement of targets requires careful planning, organization, training and follow-up. The basic steps are to:

1 Carry out a water use audit in the hotel, which will show how and where water is consumed and identifies potential areas of savings.
2 Differentiate between consumption of domestic hot water, hot water for guest use, potable cold water supplies and non-potable supplies.
3 Compare total and individual consumption figures with hotel industry benchmarks to determine potential savings.
4 Prepare a summary of opportunities.
5 Seek the advice of independent experts for analysis, recommendations and evaluation.
6 Using the water use audit results, establish realistic goals for each department in the hotel.
7 Communicate to all employees the commitment to water management and explain the objectives and goals together with data on consumption, costs and trends.
8 Gain the participation of all staff through the capitalization on their knowledge, experience and knowledge of the building.
9 Encourage staff to put forward their ideas and proposals to save water.
10 Establish a monitoring and targeting system.
11 Provide training so that staff will understand the reasons for water management and what they can do to make prudent use of utilities and to operate equipment in an efficient manner.

Who benefits from water management?

Hotel owners and managers and managers benefit, because an efficiently run building requires fewer staff and results in lower operating expenses. Reduced costs can release valuable resources that can be better employed in improving or expanding hotel facilities.

Guests benefit because an efficiently controlled hotel satisfies the needs of the guest. This in turn may result in a higher level of repeat business. Staff benefit through their empowerment, involvement and higher morale. This can lead to higher productivity, greater job satisfaction, lower levels of absenteeism and lower rates of staff turnover. The environment benefits because a reduction in the use of water resources and control of water based effluent will benefit the environment through lower consumption and decreased down-stream pollution.

By conserving all water supplies, there is a compatibility of economic interests with ecological requirements. Saving water charges, associated with unnecessary use of water together with effluent charges resulting from the disposal of waste water, saves the hotel money and conserves the water resource. Particularly for hotels located in ecologically sensitive areas, those involved with designing and managing the hotel should ensure:

- The absolute protection of the groundwater
- The safeguarding of drinking water quality
- The protection of the aquatic cycle from harmful substances.

All operators should ensure compatability with the environment of water abstraction, use and sewage treatment. While effective legal and regulatory protection of water, rivers, lakes and oceans is essential, hotel companies may need to operate with standards that are higher than the often minimal regulatory framework.

Case studies

Ramada International

Ramada International has begun to replace chlorine bleach in its swimming pools with a non-toxic ionization process. In Hong Kong salt water is used to fill toilet tanks, and sewage and laundry water are recycled for irrigation purposes.

Regency Inter-Continental, Bahrain

Since January 1992 the Regency Inter-Continental Hotel in Bahrain has been addressing environmental issues in the laundry. Bearing in mind the 'Life Cycle Analysis' – a way of looking at the environmental impact of a product right through from its ingredients, manufacture and packaging to its use and final disposal – these changes have included:

- Substituting chlorine bleaches with oxygen bleaches based on perborates, which are harmless to the environment, and degrade without affecting marine life
- Using non-phosphate, fully biodegradable liquid detergents instead of powdered phosphate detergents

- Replacing the previously used manual washing machine with a separate hydro-extractor by a new microprocessor-controlled, divided cylinder, hydro-cushion washer extractor, thereby obtaining a saving in power, water, detergents and labour.

Royal Orchid Sheraton Hotel and Towers, Bangkok

The Royal Orchid Sheraton Hotel and Towers in Bangkok is one of the luxury hotels on the banks of the Chao Phya River. The Chao Phya has become heavily polluted, owing to the rapid expansion of industry, which pumps untreated waste water into the river. The hotel has invested a large amount of money to modernize its waste water treatment unit, so that it can treat every single drop of water before it is discharged into the river. The hotel also wishes to express a positive attitude among the local community.

The 24-hour operation is manned by a group of engineers, working a split shift, who examine the quality of treated water before discharge. Carrying out such a system is expensive, but the hotel believes it creates a positive image and will help support the growth of the business in the future. The hotel has received some noticeable feedback from the public for its efforts in the form of an environmental award and the fact that the Thai media has given coverage on the waste-water system, including a well-known environmental television programme called *Lok Suay Duay Mue Rao* (or *Beautify the World with Our Hands*).

Hayman Island Great Barrier Reef Resort

The Hayman Island Great Barrier Reef Resort was redeveloped to create a luxury hotel within the context of the delicate physical environment of the Barrier Reef. The resort has been designed to have minimal visual impact, and extensive landscaping has been undertaken to blend the resort into the island's topography. A water-desalination plant produces all the island's water needs from seawater blended with rainwater collected from the island's roofs. The desalination plant is operated by means of waste heat from the power station.

The gardens are irrigated with treated effluent from the resort's sewage treatment plant, which eliminates the discharge of effluent water into the sea. Dried sewage sludge is used as a mulch and compost, as is kitchen and garden waste.

London Hilton

By the simple expedient of putting a brick into the cistern of each bathroom, the London Hilton saves an estimated 1.34 litres of water per flush.

References and further reading

Department of the Environment (1992). *The UK Environment*, Chapters 6 and 7, London: HMSO.

IHEI (1993). *Environmental Management for Hotels*, Oxford: Butterworth-Heinemann.

Lawson, F. (1976). *Hotels, Motels and Condominiums: Design, Planning and Maintenance*, Chapter 11, London: Architectural Press.

Redlin, M. H. and Stipanuk, D. M. (1987). *Managing Hospitality Engineering Systems*, East Lansing: Educational Institute of American Hotel and Motel Association.

Rose, C. (1991). *The Dirty Man of Europe*, Chapter 3, London: Simon & Schuster.

4 Energy management

The principles of energy management

One area of environmental management which has been on the agenda of hotel management for a long time is that of energy management. Unprecedented rises in the cost of oil in 1973–1974 required an immediate response in terms of energy conservation and a review of the type of energy source used. As the cost of energy increased during the 1970s, 1980s and 1990s, the majority of hotels adopted policies for the management of energy. Hotels consume more energy (expressed as £/m^2) than industrial buildings, naturally ventilated offices and secondary schools, according to the Energy Efficiency Office (1994a) and there is often considerable scope for making savings.

In addition to the interest of the hotel companies themselves, there has also an incentive from bodies such as the Energy Efficiency Office (funded by the Department of the Environment). This has been achieved through the funding of demonstration projects, the production of target consumption figures for different categories of hotels and the development of case studies based on good practice. They have shown that it should be possible to produce 5 per cent savings in energy costs through good housekeeping measures and 10 per cent savings with low-cost measures.

All parties can benefit from planned energy management. The hotel owners and managers benefit because efficiently run buildings require fewer staff and result in reduced operating expenses. These reduced costs can release valuable resources that can be better employed in improving or expanding hotel facilities. Guests gain because an efficiently controlled hotel satisfies their needs at a lower cost, in turn helping the business in terms of a higher level of repeat business. Staff benefit through increased job satisfaction, lower rates of absenteeism and lower staff turnover. This increases productivity which can result in financial savings that can be passed on to the guests.

Finally, there are obvious benefits to the environment. A reduction in the use of non-renewable energy resources conserves the energy supply and also diminishes some of the negative impacts resulting from the use of fossil fuels, including atmospheric pollution, global warming, ozone depletion and acid rain.

The principles of energy management are to reduce the amount and cost of energy used by the hotel, subject to the constraint that at no stage should there be a perceived loss of comfort level supplied to the guests unless this is done with their consent. As an example, it may be possible to reduce energy costs by allowing the guests to choose whether or not they need fresh towels every day.

Energy can be saved by a series of measures including:

● A review of the mix of energy sources used
● A review of tariffs used or other contractual arrangement with energy supply companies
● Staff training leading to practical steps that can be taken to reduce energy consumption
● A programme of capital investment on the building, plant and equipment in order to reduce energy consumption.

Before any programme of changes is commenced, an energy policy should be developed, based on an initial audit of the current position. This should be followed by a planned series of projects, targeting areas of high wastage. Regular energy audits should be conducted in order to review progress and to revise the plan.

Energy supplies

Forms of energy

There are a number of different forms of energy (Kirk and Milson, 1982):

1 Mechanical energy exists in the form of kinetic energy (the energy in a moving body, such as a pendulum), potential energy (the energy in a body because of its height, such as a gravity feed water tank), strain energy (such as in a spring) and pressure energy in a gas or liquid.
2 Nuclear energy is that energy contained in the nucleus of an atom, which can be liberated either by splitting the atom (nuclear fission) or by combining atoms together (nuclear fusion).
3 Electrical energy is associated with the distribution of electrical charges.
4 Electromagnetic energy is contained in radiation from the sun, light waves, microwaves and radio waves.
5 Thermal energy (or heat) is associated with the vibration of molecules caused by an increase in temperature.
6 Chemical energy is stored in a substance because of the intramolecular forces holding the molecules of the substance together.

The term 'fuel' represents a particular form of chemical energy associated with carbon rich materials which, when combusted in the presence of oxygen, release large amounts of thermal energy, together with carbon dioxide, water and other chemicals depending upon the precise composition of the fuel. The process of combustion is very important to all human activity since it is used directly to heat buildings and water, to cook food and indirectly to generate electricity.

Chapter 1 defined the difference between renewable and non-renewable forms of energy. Currently, in the UK, 98 per cent of all energy comes from non-renewable sources (including 22 per cent from nuclear energy) and only 2 per cent from renewable sources although this proportion is set to increase to 5 per cent by the year 2000 (HMSO, 1994).

Units of energy

Energy is defined as 'the ability to do work' and it therefore has the same unit as 'work' which, in the SI system of units, is the joule (J). However, because the joule is rather a small quantity of energy, it is usually found in multiples, such as:

kilojoule (kJ) =1000 J
Megajoule (MJ) = 1 million J

The term 'power' is used to describe the rate at which energy is used, i.e. the amount of energy consumed in a period of time. The SI unit of power is the watt (W), which is the consumption of energy at the rate of one joule per second. As with the joule, the watt is rather a small unit and therefore multiples are often used:

kilowatt (kW) = 1000 W
Megawatt (MW) = 1 million W

The kilowatt also gives us an alternative unit of energy, one which is often used in energy management literature, the kilowatt-hour (kWh). This is useful because it is a very practical measure. Assuming that we have an appliance which has a power rating of 1 kW, if it is in constant use for 1 hour, the appliance will consume 1 kWh of energy. Energy consumption in kWh is given by the equation:

Energy consumption (kWh) = power rating of appliance (kW) × time in use (h)

For example, if a water boiler, with a power rating of 5 kW, takes 10 minutes to boil a tank of water, its energy consumption is:

5 . 10/60 = 0.833 kWh

Care must be taken when using this relationship in the case of appliances which are controlled by a thermostat, since the appliance will go through an on/off cycle under the control of the thermostat.

Sometimes it is necessary to convert from kilowatt-hours into joules and vice versa. This is relatively straightforward. One kilowatt-hour is the equivalent of 1 kilowatt of power consumed for 1 hour, which is 3600 seconds. Therefore:

1 kWh = 3 600 kJ or 3.6 MJ.

Energy conversions and energy efficiency

The first law of thermodynamics tells us that energy cannot be created, nor can it be destroyed. However, it is possible to convert from one form of energy into another, but each time a conversion takes place some of the energy may be effectively lost from the system in the form of low-grade heat, i.e. heat at a low temperature which escapes to the environment. For example, a water boiler may be used to convert chemical energy from a natural gas supply into thermal energy, in the form of hot water. However, some energy will be lost from the system, since the products of combustion will be discharged from the flue of the appliance and these combustion products will contain some heat. We can express the effectiveness of the conversion process in terms of its efficiency, which is the ratio of useful energy output to total energy input (see Figure 4.1). If the energy input to the boiler is 5 MJ and the hot water generated is equivalent to 4 MJ, then the efficiency of the appliance is 4/5 or 80 per cent.

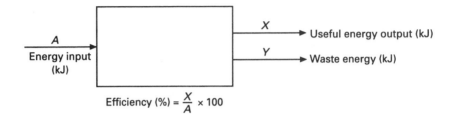

Figure 4.1   *Efficiency of an energy conversion*

Some energy conversions are less efficient than others (see Table 4.1). For example, when thermal energy (from a fossil fuel) is converted into electricity the efficiency may be only as high 36 per cent, with the remaining 64 per cent being lost as waste heat at the power station. However, if we can reuse this waste heat for other purposes, such as heating a greenhouse or providing a supply of hot water, then the overall efficiency can be increased. This is the theory behind district heating schemes and combined heat and power systems.

Table 4.1   *Examples of energy conversion efficiencies*

Conversion	Example	Efficiency (%)
Chemical to heat	Gas water boiler	70–90
Heat to electrical	Hydroelectric	90
Electrical to mechanical	Electric motor	70–90
Heat to mechanical	Steam turbine	45

When carrying out energy audits, it is often necessary to convert quantities of fuels into units of energy and to perform conversions from one unit to another. In this situation the data given in Table 4.2 should prove of value. These are input values. In order to obtain the output of energy, the input value must be multiplied by the efficiency of the appliance. Precise energy values of fossil fuels vary depending upon the chemical composition. Accurate data can be obtained from suppliers.

Table 4.2   *Energy conversion data*

Natural gas	$m^3 \times 10.6$	=	kWh
	$ft^3 \times 0.3$	=	kWh
	therms $\times$ 29.3	=	kWh
LPG (propane)	$m^3 \times 25$	=	kWh
Coal	kg $\times$ 8.05	=	kWh
Coke	kg $\times$ 10.0	=	kWh
Gas oil	litres $\times$ 12.5	=	kWh
Light fuel oil	litres $\times$ 12.9	=	kWh
Medium fuel oil	litres $\times$ 13.1	=	kWh
Heavy fuel oil	litres $\times$ 13.3	=	kWh

All fossil fuels, when burned to release energy, also produce CO_2, as shown in Table 4.3. (Energy Efficiency Office, 1993). It is estimated by the Energy Efficiency Office that, in one year, a typical hotel releases about 160 kg of CO_2 into the atmosphere for every m^2 of floor area. The figure for electricity relates to the coal, oil or gas utilized to make the electricity at the power station. Clearly, if the electricity comes from nuclear energy or a hydroelectric plant, the figure for CO_2 will be zero.

Table 4.3   *Release of CO_2 associated with the use of energy*

Fuel	CO_2 per kWh
Gas	0.21
Oil	0.29
Electricity	0.72

Secondary energy sources

A secondary energy is one where the energy has undergone conversion from one form to another. The most common secondary energy source is electricity where thermal energy, from fossil fuels, and kinetic energy, in moving water, are converted into electrical energy. The process of converting fossil fuels into electricity has a low efficiency, but the use of combined heat and power (CHP) can increase this to 80-90 per cent. Much of the development of CHP has been in the commercial sector.

Non-renewable energy supplies

Currently the most abundant source of energy comes from the fossil fuels – coal, oil and gas (see Figure 4.2). Recently there has been a switch from oil-based fuel oils and coal to natural gas in terms of both industry and domestic use. Coal and solid fuels are still used for the generation of electricity, but even here there has been a switch to natural gas. This is partly on the basis of economic factors but also because natural gas supplies do not produce the same levels of SO_2 in the flue gas. However, coal stocks are sufficient to ensure supplies well into the next millennium. Coal is recovered from the ground by either opencast or underground mining. There are a number of different types of naturally occurring coals – peat, brown coal, lignite, bituminous coal and anthracite. In addition to these there is also a wide range of manufactured smokeless fuels.

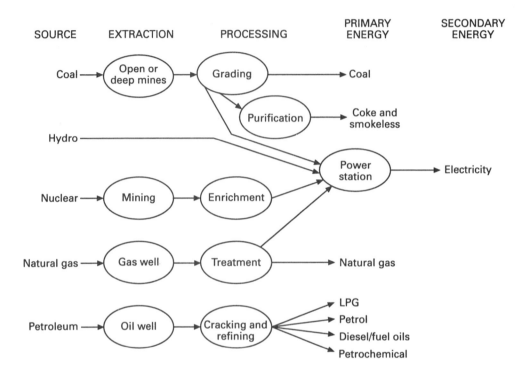

Figure 4.2   *Major primary and secondary sources of supply*

Fuel oils are derived from petroleum during the refining process. Worldwide the extraction of petroleum will reach its peak around the year 2000, and by 2050 will be almost depleted. Additional petroleum type products will become available from tar sands and oil shale. Alternatively, petroleum can be replaced by wood alcohol for use in internal combustion engines.

When petroleum is refined, in addition to petrol and diesel, a number of fuel oils are produced from light (domestic) fuel oils through to heavy fuel oils. The latter are more economic for use in large hotels but the initial cost of the plant to burn heavy fuel oils is more expensive. In addition to viscosity, fuel oils vary in their sulphur content.

Natural gas consists predominantly of methane (CH_4). It is delivered by pipeline direct to the property. Particularly in remote locations where the natural gas distribution is not available, commercial propane (Calor gas) is an alternative. This is made up of 82 per cent propane (C_3H_8). Other than for transport purposes, natural gas is now the major source of fossil fuel, with 65 per cent of domestic fuel use, 42 per cent in commerce and 32 per cent in industry (HMSO, 1994).

Nuclear energy, also based on a non-renewable energy source, is used for the generation of electricity. In the UK, approximately 20 per cent of electricity comes from this source. The main fuel for nuclear reactors is uranium-235.

Renewable energy supplies

Most renewable energy supplies derive from solar energy – wind, hydroelectric and wave together with solar power. Other possibilities are tidal power (from the movement of the moon around the earth) and geothermal energy stored in the earth's crust (see Table 4.4). Currently, only a small proportion of the world's energy supplies comes from these renewable sources.

Table 4.4 *Renewable energy sources*

Primary		Secondary
Solar		Thermal
Solar		Electrical
Biological	Wood	Thermal/electrical
	Organic waste	Thermal/electrical
Water	Water mill	Mechanical
	Water turbine	Electrical
Wind	Windmill	Mechanical
	Wind turbine	Electrical
Tidal	Generator	Electrical
Geothermal		Thermal

The energy management programme

The role of the energy manager

A fundamental principle of energy management is that someone should have a responsibility for the energy use within the hotel or group of hotels. In a large group of hotels, there may be an individual who has the specific title 'energy manager', but more usually it will be the person responsible for the building and its maintenance and who may have responsibility for the total environmental management programme.

The energy audit

The energy audit is an essential step in the establishment of a professional energy management programme. The objective is to analyse and evaluate collected data to determine the energy performance of the entire building. This is relatively easy for the whole hotel, but requires more planning and investment if it is to be done for departments or major consumers.

Energy consumption for the whole hotel can be obtained from historic records of utility consumption data, relevant hotel statistics, equipment technical data and weather information.

An energy audit should be conducted, which will show:

● Energy consumption and cost data for up to five years
● Frequent meter readings to show day-time, night-time and weekend energy consumption
● An inventory of all energy-consuming equipment showing age, power loading and maintenance record, together with data on frequency of use.

Particularly for expensive energy use areas such as kitchens and the laundry, sub-metering can be an important first stage in reducing energy waste since it allows the monitoring of energy consumption down to a level at which control can be exercised. It also allows for the full implementation of energy monitoring and targeting programmes. The installation of sub-meters for each of the utilities may be of value in order to assess consumption by major energy consumers, such as chillers, boilers, air-handlers and the kitchen. Although this is an expensive process, it is well worth doing, as lack of information on consumption within individual departments is often the cause of failure of savings programmes. It has the following advantages:

● Accurate energy audits
● Correct determination of true efficiencies of major consumers such as chillers, boilers and air handlers
● The ability to trace inefficiency and waste
● Immediate feedback of results of specific energy conservation measures that would otherwise be lost in the total consumption of the building
● Enhanced departmental management accountability; by placing responsibility with the individuals who control and consume utilities
● Determination of the feasibility of capital investment projects and their resultant true savings after installation
● Control of public utility meters and deliveries
● Information on consumption for designers and energy experts for the proper sizing of new or replacement equipment.

Assessing current performance

By collecting energy consumption data, it is possible to monitor changes over time and to compute a performance indicator for the hotel which can be compared with national data (Energy Efficiency Office, 1990, 1993). The energy audit provides a detailed evaluation of energy efficiency, but it has little meaning without comparison. When audits have been carried out for a number of years, a comparison can be of year-on-year performance. Comparison with other hotels can also be an important yardstick.

According to the Energy Efficiency Office (1993), energy consumption in hotels depends upon the type of hotel, together with its size, method of construction, climatic factors and geographic location. In particular, the mix of energy sources used in the hotel will play an important part in approaches to energy management. Typical energy sources, based on average value for 100 international hotels of between 200 and 1000 bedrooms, are shown in Figure 4.3. They suggest figures for good, fair and poor energy use in three types of hotels in the UK, as shown in Table 4.5. Care must be taken in interpreting these figures as there are many constructional, location and operational factors which can influence energy consumption. Does the hotel have its own laundry or is this work contracted out? Does it have a leisure centre or swimming pool? In particular, air-conditioning will increase the consumption of electricity by about 50 per cent and the fossil fuel consumption by about 10 per cent.

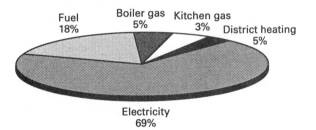

Figure 4.3   *Energy consumption based on 100 hotels of size 200–1000 rooms (based on IHEI, 1993, Figure 3.1)*

Table 4.5   *Typical energy consumption for hotels (kWh/m²)*

Category	Good		Fair		Poor	
Energy source	Fossil	Electricity	Fossil	Electricity	Fossil	Electricity
Luxury hotel >150 rooms	<300	<90	300–460	90–150	>460	
Business/holiday hotel >140 rooms	<260	<80	260–400	80–140	>400	
Smaller hotel >120 rooms	<240	<80	240–360	80–120	>360	

Based on Energy Efficiency Office (1993).

In order to calculate total energy consumption and to compare this with industry-average figures it is necessary to convert all amounts of energy and fuel consumption to the same units of energy. This is commonly expressed as kWh. It is not a simple matter to convert these delivered energy figures into costs, because each source of energy has a different price per kWh and a different efficiency. Before energy management can be implemented in any hotel, an accurate estimate of these data is required. It is possible to obtain conversion factors from suppliers of fuels so that energy input values can be calculated. (See Table 4.2 for approximate values.)

The allocation of energy consumption to the various activities which take place in a hotel will vary from one hotel to another, but typical figures for the UK are shown in Table 4.6 and in diagrammatic form in Figure 4.4. Because of the different costs and efficiencies associated with each form of energy, Figure 4.5, which shows the breakdown by costs gives a quite different picture.

Table 4.6   *Annual delivered energy consumption of a typical hotel*

Use	Energy kWh/m^2 Fossil fuel	Electricity	% of total
Heating	225	9	47.3
Hot water	96	3	20.0
Catering	56	17	14.7
Lighting	0	40	8.1
Other	7	42	9.9
Total	384	11	

Based on Energy Efficiency Office (1993).

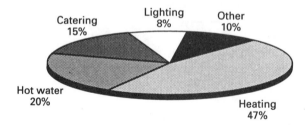

Figure 4.4   *Delivered energy by use (based on Energy Efficiency Office, 1993)*

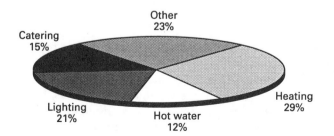

Figure 4.5   *Delivered energy by cost (based on Energy Efficiency Office, 1993)*

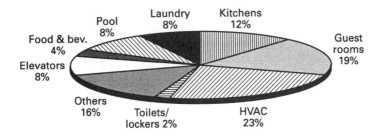

Figure 4.6 *A typical energy audit (based on IHEI, 1993, Figure 3.2(a))*

More detailed figures based on the energy audit for a 200-bedroom hotel which was fully metered are shown in Figure 4.6. There will be specific aspects of any hotel's operation which will distort the actual energy consumption (for example, if a hotel has an in-house laundry, indoor pool, health club or air-conditioning). The consumption figures are also affected by climate and by occupancy rates. Gross figures, as calculated above, can be compared to industry averages by applying correction factors (IHEI, 1993, p. 34).

Changes in occupancy affect water and energy consumption considerably. We might expect that there would be a linear relationship between occupancy and consumption since as the number of occupants increases, more energy and water will be required. However, there is a base load of consumption which will occur even if there are no guests in the hotel (see Figure 4.7).

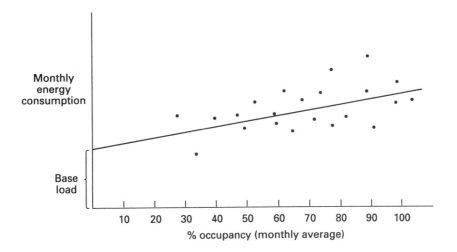

Figure 4.7 *The relationship between energy and occupancy*

Where boilers are used for both heating and domestic hot water supplies, it is often difficult to determine the precise breakdown. In general, the heating energy can be assumed to be 60 per cent of total fossil fuel consumption (Energy Efficiency Office, 1990).

One way of obtaining a more precise estimate is through plotting fossil fuel consumption for each month. The non-heating load (which occurs during the non-heating season) is a measure of consumption related to hot water utilization. However, this assumes a number of factors, such as a similar distribution of occupancy levels across summer and winter months and the fact that guests use similar amounts of hot water in summer and winter (see Figure 4.8).

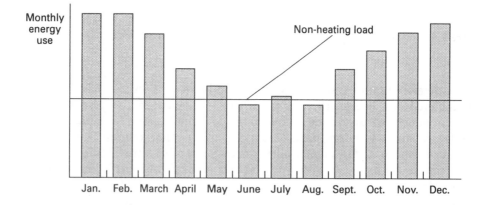

Figure 4.8 *The relationship between energy use and heating season*

Degree Days

Climate is one factor which has a considerable effect on energy consumption. In cold climates the heating system will be required to operate for longer periods and against lower external temperatures. In hot humid climates, air-conditioning will be needed to reduce temperature and humidity. These differences can be expressed in the form of *Degree Days*.

If we consider heating requirements, it is useful to have a measure of how much heating is required based on a combination of outside temperature and length of time. This can be done by first agreeing on an external temperature below which heating is required. This varies from one country to another. For example:

In Germany 15°C
In the UK 15.5°C
In the USA 18.3°C

By obtaining the average outside temperature for each day and subtracting this from the agreed threshold temperature we can calculate the number of degrees of heating required for that day. If we add together all these degree days for January we get the total number of Heating Degree Days for that month. These data are supplied in the form of 'Degree Days' and are published on a regular basis for different areas of the UK in journals such as *Energy Management* (examples of which are shown in Table 4.7). A similar approach can be used for Cooling Degree Days.

Table 4.7 *Typical Degree Day data for the UK*

Month	Thames Valley	North-east Scotland
May 1994	116	217
May 1993	92	205
May (20-year average)	114	198
April 1994	198	262
April 1993	153	246
April (20-year average)	199	270

In winter, we are concerned with the amount of energy which must be supplied to the building in order to compensate for heat losses to the outside environment. A Degree Day is the difference, over a 24-hour period, between the average external temperature and a defined base temperature. The individual figure for each day is added together to give the total Degree Days for that month. As an example, for the UK, the base temperature is 15.5°C. If the mean outside temperature for a 24-hour period is 10°C, then that day contributes 5.5 (15.5–10) Degree Days to the monthly total. This can then be related to the energy consumption for that month in order to determine its contribution to the heating load.

Energy consumption data are also affected by exposure to wind and rain. A hotel in an exposed site can have a space heating requirement which is 20 per cent more than an equivalent building in a sheltered site.

In periods of warm weather, when air-conditioning is used, we can use Degree Days to determine their contribution to the cooling load. In this case we are concerned with the amount of energy which must be supplied to the building in order to compensate for heat gain from the outside environment.

The collection of Degree Day data and heating fuel consumption data, where the heating system has separate metering, can provide useful information. If the Degree

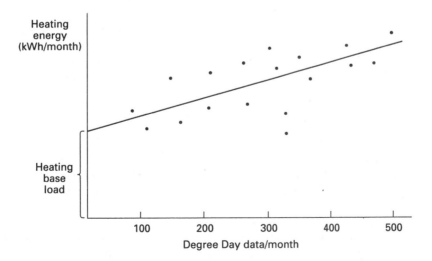

Figure 4.9 *The relationship between Degree Day data and heating energy consumption*

Day data and the fuel consumption data are plotted over a period of time it allows analysis of the relationship between heating energy and the heating load (see Figure 4.9). Careful consideration of this graph can be useful in assessing possible problem areas and potential for savings. From this graph, we can identify two important characteristics: the heating base load and the slope. The heating base load, which is the intersection of the regression line with the vertical axis, shows the level of heating consumption in the building for a Degree Day figure of zero. A high base load might indicate ineffective manual and automatic controls, such as defective thermostats or thermostat settings which are too high. A high slope to the line indicates poor building thermal performance and perhaps the need for better insulation.

Project appraisal

All capital projects require a full financial appraisal as well as the justification from the point of view of the environmental impact. Where outside contractors or consultants are being used, it is useful to ask the following questions about the organization:

- Is it of good standing?
- Does it have experience of working for hotel companies?
- Can the proposed savings be justified against data from the audit – are projected savings realistic?

A financial appraisal should identify the following:

- What is the capital cost of the project?
- What are the projected annual savings on utilities?
- What is the effect on maintenance costs?
- What is the expected lifetime of the installation?

There are alternative methods of capital project appraisal, from a simple payback method through to return on investment projections.

Many hotels are sub-contracting their energy management to a contract company. In addition to managing energy consumption on a day-to-day basis, these companies will often fund capital projects on the basis of taking a proportion of any savings (Energy Efficiency Office, 1994b). They also allow the hotel company to concentrate on its core business.

Energy-conservation measures

At each stage in the life-cycle of a hotel, there are opportunities for adopting energy-conservation measures. For example, at the design stage of a new hotel there are many opportunities for building in the efficient use of all utilities. When new buildings are being designed it is important to ensure that figures for lighting, heating, cooling and total load requirements in kWh per square metre comply with what the state-of-the-art building techniques can provide. Similar opportunities may arise during major refurbishment projects. As another example, if solar gain is considered at the design stage, simple passive solar gain measures can contribute to the heating of a hotel and, depending on climatic conditions, may result in a 15 per cent energy saving for space heating.

At the time of commissioning, all the systems in the building should be checked to make sure that they are working in accordance with the design specifications. Test certificates should be obtained which confirm the completion of these tests. At the time a building is handed over to the hotel company there are many pressures to start business as soon as possible in order to generate revenue. Experience shows that some jobs never get satisfactorily completed nor specific problems resolved. If these problems are not addressed at the time of commissioning and hand-over, the hotel will be penalized with complaints about comfort and higher utility rates for the rest of its life.

However, the above opportunities only occur at rare intervals in the life of a hotel and there is considerable scope for introducing changes during its normal working life. There should be a planned programme of energy-conservation measures. It is logical to start with those actions which will give the greatest savings at the least cost. The success of these actions will provide positive feedback on the programme as a whole and will help to justify the capital investment on more expensive actions. All proposed changes should be evaluated, according to the decision-making flow chart shown in Figure 4.10.

The types of actions taken can be grouped according to a hierarchy as follows:

1 Very low: investment in a programme of staff-awareness training followed by specific training measures in areas of good housekeeping
2 Low cost: match source to load by shutting off equipment not required, by ensuring that controllers, timers and programmers are correctly set for the climate and for the level of activity of the hotel
3 Medium cost: ensure that all equipment is operating at maximum efficiency through regular maintenance of heating equipment, chillers and pumps and through the replacement of obsolete control equipment with microprocessor-based controls
4 High cost: invest in heat-recovery systems where appropriate
5 Very high cost: invest in alternative energy sources, such as combined heat and power (co-generation), solar energy, wind and water energy, geothermal energy, energy from burning waste.

The plan should be to start where possible with low-cost measures, together with obvious areas of waste and to target activities which can be certain of quick results in order to build confidence in the programme. Only when this confidence, and the resulting enthusiasm, is shared by the majority of the team should more costly measures be attempted.

Some measures carry no costs. For example, the evaluation of utility tariffs and energy contracts should enable the organization to ensure that it is obtaining the best value for money. Utility companies may offer a number of different types of contract and this should be optimized to give the lowest cost for the company based on the total energy consumption and the pattern of use. In the case of supplies such as gas and fuel oil, a number of different companies should be invited to tender for the provision of energy.

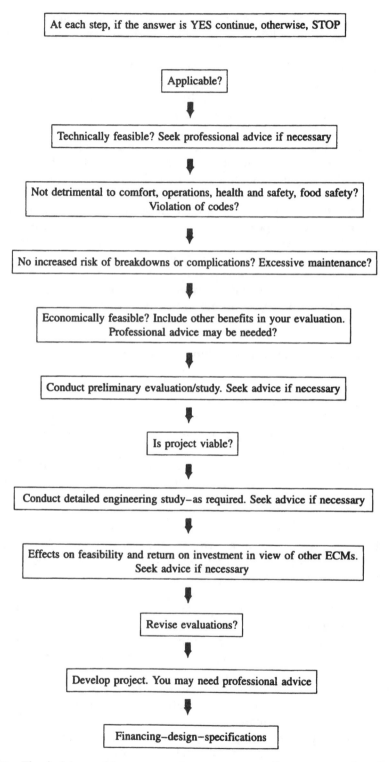

Figure 4.10 *The decision-making process of energy-conservation measures (based on IHEI, 1993)*

In the case of electricity, the hotel may be charged a Maximum Demand Tariff, which is based on the peak loading at any time within the charge period. It is designed by the supply company to discourage users from having large peaks and troughs in their demands for electricity. Hotels which are subject to this charge pay a variable charge based on this peak demand. Where this is the case, the hotel should investigate causes for the peak loading. If several large electrical appliances are on at the same time, this may lead to a peak of demand. If this is the case it may be possible, by scheduling the use of these appliances, to reduce the load – a technique known as manual load shedding. For large hotels which are paying a large maximum demand charge it is possible to fit an automatic load-shedding control. This is an electronic device which constantly monitors electricity consumption and, when this nears a pre-set peak value, automatically switches off appliances, based on a pre-programmed priority. The priority is based on the fact that some uses of electricity can be delayed without compromising the safety and comfort of guests and employees.

Simple low-cost measures can produce immediate savings on energy costs. The only requirement for many of these changes is a programme of energy management awareness seminars and specific staff training programmes. For example, all staff should be trained to switch off appliances and lighting when they are found to be on but not in use. Maintenance staff should regularly check on thermostats, programmers and time switches to ensure correct settings. Staff should be encouraged to report faults, such as dripping taps and faulty controls. The use of hot water by housekeeping staff should be reviewed to ensure that procedures are consistent with standards set during training. Other low-cost measures include regular maintenance of all energy-consuming appliances.

Monitoring and targeting

The achievement of targets requires careful planning, organization, training and follow-up. The basic steps are:

1 Carry out an energy audit in the hotel, which will show how and where energy is consumed and identifies potential areas of savings.
2 Compare total and individual consumption figures with hotel industry benchmarks to determine potential savings.
3 Prepare a summary of opportunities.
4 Seek the advice of independent experts for analysis, recommendations and evaluation.
5 Using the energy audit results, establish realistic goals for each department in the hotel.
6 Communicate to all employees the commitment to utility management and explain the objectives and goals together with data on consumption, costs and trends.
7 Appoint an energy co-ordinator (usually the engineer), define responsibility within each department and develop an effective communication system.
8 Gain the participation of all staff through the capitalizing on their understanding, experience and knowledge of the building.
9 Encourage staff to put forward their ideas and proposals to save energy.
10 Establish a monitoring and targeting system.

11 Provide training so that staff will understand the reasons for energy management and what they can do to make prudent use of utilities and to operate equipment in an energy-efficient manner.
12 Provide continued motivation through developing standard operating procedures, efficiency charts and other regular communications.
13 Review current supply arrangements frequently with existing supply companies to ensure that optimum tariffs are in use.

Some basic rules for good practice include:

● Provide a comfortable environment for customers.
● Improve efficiency by training all personnel to understand, operate and maintain the hotel's equipment and systems in an energy-efficient manner and to minimize the waste of energy.
● Invest in the building, equipment and systems to improve efficiency.
● Measure efficiency as a standard procedure, particularly for those areas that are major consumers such as boilers, chillers, cooling towers and air-handlers.
● Match utility use to the demand which is a factor of business activity (time of day, day of week and occupancy) and of climate.
● Establish profit centres for energy use, through the establishment of sub-meters and the allocation of costs to each department based on their actual energy consumption.
● Place utility-conservation projects on the same level as projects related to interior decoration, structural changes, extensions, additions etc.
● Good training is the first and most important step and it should be an on-going process for all staff.

Guidelines for major use areas

Heating, ventilation and air conditioning

This area usually represents the greatest single consumer of energy in the hotel, representing somewhere between 25 per cent and 50 per cent of the total energy cost. The lower figure is applicable to smaller unsophisticated hotels in moderate climates and the higher one to luxury hotels in subtropical and tropical countries. The potential for energy efficiency is in most cases very high. Opportunities fall into the following major categories:

● *Building characteristics* energy-conservation factors designed into fabric; retrofitting energy conservation measures;
● *Matching source to load* flexible systems which are efficient across the range of operating conditions (number of customers, climate etc.); heating systems may be 70–80 per cent efficient at full-load but only 30 per cent efficient at one-third load. The hotel may operate at full load for only a few days a year, for example in the most extreme climatic conditions and with full occupancy. Sophisticated control systems, linked to computer logging facilities, can optimize heating systems
● *Decreasing loads* reducing lighting consumption, decreasing number of air changes, peak demand control

- *Increasing efficiency* improving efficiency of plant across the range of operating conditions
- *Reducing costs* cheapest utility source and best available tariff
- *Recovery of waste energy* recover the energy normally lost to the atmosphere or drains from exhaust systems, swimming pools, condense heat recovery, boiler flue heat recovery
- *Use of alternative energy* solar energy, co-generation, refuse burning
- *Sources* geothermal energy, wind energy, water energy.

The fabric of the building, particularly in the case of old ones, should be investigated to assess the extent of heat loss through walls, windows, roof and draughts. If these are excessive, the potential of energy savings should be investigated. A quick and cheap way of doing this is by monitoring the rate of snow and frost melt in winter months.

The temperature of all public areas and guest rooms should be investigated, to ensure that these are not set at too high a temperature. Simple measures, like the replacement of conventional radiator valves with Thermostatic Radiator Valves (TVRs), can save money and allow guests to adjust the room temperature according to their personal preference. In hotels there may be a build-up of heat in upper floors due to lights and heating, exacerbated by the fact that warm air rises.

Old boilers will generally have lower efficiencies than their newer equivalents and, in addition, are more suitable for sophisticated control. Audits of existing hotels should include measures of boiler efficiency in order to determine any potential for savings. For small hotels, a condensing boiler is more efficient in use than a conventional one. In the condensing boiler heat is extracted from the water vapour which is entrained in the flue gases through the use of extra heat exchange area. The latent heat trapped in this moisture vapour can represent 10 per cent of the total energy content of the fuel. Condensing boilers can increase boiler efficiency from 80 per cent up to 90 per cent.

Building energy management systems (BEMS) are sophisticated computer-based systems. BEMS can optimize boiler performance and can be used to provide sophisticated management to remote locations (see Figure 4.11) through the use of auto-dial modem systems (Kirk, 1987). In this way management support can be provided by head office personnel or through the use of contract companies. Replacement of the boiler and the use of a modern BEMS can save large amounts of energy. As an example, a BEMS, in conjunction with new high-efficiency boilers allowed the Ritz Hotel to save 40 per cent of its gas consumption and reduced its energy consumption from 526 kWh/m^2 to 322 kWh/m^2 (Energy Efficiency Office, 1994b).

BEMS can be used to control heating zones within a hotel. Each part of the hotel can be operated as a separate heating zone with each zone having its own pumped circulation. Each zone can also have its own heating programme and this can be related to operational factors, such as level of occupancy. At quiet periods, rooms can be let in a single zone, with other areas receiving only background heating. As occupancy increases, more zones can be switched from background to full heating.

Occupancy-linked controls can allow energy consumption to be targeted only on those rooms that are occupied. A number of different systems have been used (Energy Efficiency Office, 1994c). In one, a key-link panel in reception activates the system and is linked to two-stage thermostatic radiator valves (TVRs) in the guest's room. When the key is present on the rack the room temperature is held at a 'set-back' temperature.

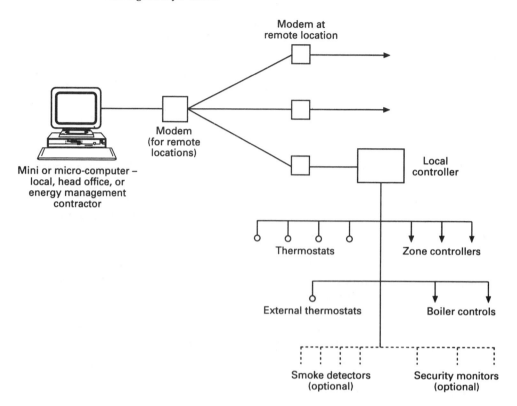

Figure 4.11   *A building energy management system*

When the key is removed from the key rack and handed to the guest, a signal is sent to the TVR, causing the setting to be increased to the normal operating temperature. The system also allows guest control of the temperature via a wall thermostat. An alternative system utilizes a key fob switch just inside the door of the guest room (see Figure 4.12). The key, which is used to open the door, is 'parked' in the key fob holder which activates electrical supplies to lights, electrical outlets for kettle and TV. This prevents lights and appliances being left on in empty rooms. It may also control room temperature as described in the case of the key rack system. A third alternative which has been used is a passive infra-red occupancy detector linked to a two-stage TVR. When there is no signal from the occupancy detector, the thermostat is set to its setback temperature. When movement is detected, the temperature is set by a room temperature control, by the guest. When movement ceases to be detected, the control holds the room at the guest set temperature for a predetermined time, which can be varied by between 5 and 60 minutes.

Showers will use less hot water than baths and flow restrictors on showers can reduce this even further. Thermostatic controls on domestic water supplies can save energy.

Combined heat and power (CHP), or co-generation as it is referred to in the USA, involve using a fuel supply (oil or gas) to drive a small generator which produces both electricity and heat. In a conventional electricity generator the heat produced is regarded as a waste product. In a CHP system, the heat contributes to normal hot

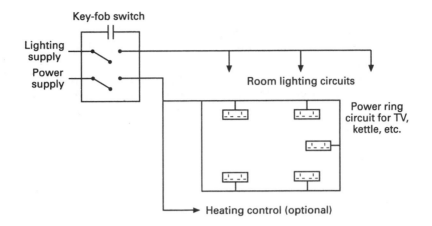

Figure 4.12   *A key fob-activated guest room control*

water and space heating requirements, giving it an overall efficiency of 80–90 per cent. Small-scale CHP systems can save energy in hotels which have a constant demand for hot water throughout the year. A single CHP unit (made from a modified car engine) can supply in the order of 40 kW of heat and about half this amount of electricity (see Figure 4.13). Single units can go up to 1000 kW of electricity and 2 kW of heat. The unit can save between 10 per cent and 25 per cent of total electricity and gas costs. It can also reduce total CO_2 emissions to the atmosphere. Larger installations use multiple CHP units in order to provide greater flexibility of output related to electrical and heating demand. A significant proportion (20 per cent) of the total CHP installations in the UK is in hotels.

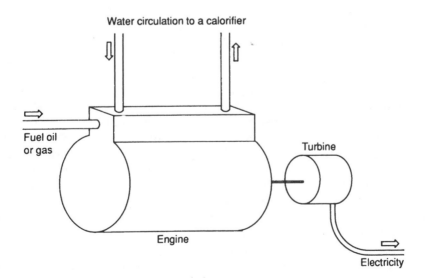

Figure 4.13   *A combined heat and power system*

Energy and financial savings are not guaranteed with the installation of CHP. Careful design of the system is required and, in particular, accurate sizing of the CHP system is required. Maintenance costs and the effect of breakdown on performance also need to be considered.

District heating may be possible for a hotel which is located close to an electricity generating plant. In a conventional electricity power generator 35–50 per cent of the energy value of the fuel is converted into electricity, with the remaining energy lost as heat. In some situations, this waste heat may be used to supply local industry, commerce or housing – this is known as District Heating. If a hotel is located in close proximity to a power station, it may be possible to purchase heat (in the form of a steam or hot water pipe distribution) which can be used to supplement the conventional energy supply.

Lighting

Lighting accounts for some 15–20 per cent of a hotel's electricity consumption. In addition, heat generated by lights increases air conditioning loads. In winter it assists the heating system, but in a very inefficient manner.

Lighting input figures are measured in watts. This is not a direct measure of output itself since the relationship between input and output depends upon the method used to convert electricity into light. The unit of light output is the lumen: a 40 W incandescent light bulb produces about 450 lumens whereas a 40 W fluorescent tube produces about 2150 lumens – nearly four times as much light for the same input.

- *Use high efficiency bulbs*   replace incandescent light bulbs with fluorescent tubes which give four times more light for a given consumption of electricity and can last up to ten times longer.
- *Match load to demand*   switch off lights when an area of the hotel is not in use and consider the installation of switches, timers, dimmers, photocells and motion detectors.
- *Improve efficiency of lighting*   regularly clean light fixtures, improve reflection from walls through the use of brighter colours and replace lightshades with translucent types.

A hotel can save significant amounts of electricity by converting from tungsten filament light bulbs to compact fluorescent and low-voltage tungsten halogen lighting. There are also reductions in maintenance costs because of the long life of these bulbs. When this was first suggested there was some resistance to the use of fluorescent lighting in public areas and guest rooms because of the utilitarian appearance of fluorescent tubes and the quality of the light. Now, with modern styles of compact fluorescent and low-voltage lighting, this concern has been overcome. Compact fluorescent bulbs can now be used even in high-profile lighting such as chandeliers. In hotels which have adopted such bulbs there has on occasion been a theft problem, which can be overcome through using non-standard light fittings.

Dimming circuits can also provide energy savings as well as allowing lighting levels to be more readily adjusted to the needs of guests (see Table 4.8). Microprocessor-controlled lighting management systems can be used to predetermine lighting levels in various parts of a building at different times of the day and year.

Table 4.8 *Running cost of artificial lighting*

Lamp type	Lumens per watt	Lamp life (hours)	Cost ratio (tungsten filament = 1)
Tungsten filament	10–15	830	1
Low-voltage tungsten halogen	25–30	10,000	8
Compact fluorescent bulbs	50–80	8,000	15
Fluorescent tubes	50–80	8,000	10

Some care is needed when replacing spotlamps with compact fluorescent lamps because the light output depends not only on the bulb but also on the type of luminaire (light fitting). A compact fluorescent bulb in the incorrect type of luminaire may not always produce the equivalent amount of light output as the spotlamp/luminaire combination.

There is more discussion of lighting as a contributor to the indoor environment in Chapter 5.

Guest rooms

Guest rooms consume a major proportion of the hotel's energy and water, typically 30 per cent of the total for the hotel. Consumption is governed by two major factors: climate and occupancy levels. Climate affects heating and air-conditioning loads, occupancy affects other sources of energy and water consumption. Savings can be achieved by:

Monitoring utility consumption on an hourly basis for 24 hours and noting the relationship between usage and activity. Excessive use of water during the night might indicate leaks, and high usage of water between 10 am and 4 pm might indicate excessive usage by housekeeping staff for cleaning purposes.

Modifying the cleaning procedures of housekeeping staff. In some hotels cleaning of bedrooms can account for one-third of water consumption in the rooms.

During periods of low occupancy, concentrating rooms allocated to areas of the hotel which are zoned for heating purposes. This allows those areas of the hotel not in use to be shut off.

Adjusting thermostats for summer and winter use to prevent guests adjusting temperatures by opening windows.

Training housekeeping staff to switch off lighting and televisions as soon as rooms are vacated.

During hot or cold weather keeping curtains closed to reduce heating/cooling gains.

Adjusting the volume of water usage by toilet flush to 6–8 litres.

Installing pressure regulators on shower-heads and flow restrictors on water taps and sinks.

Installing thermostatic control valves on radiators.

Installing key switches on power supplies to rooms.

Kitchens

Kitchens are traditionally among the least energy efficient operations in hotels and catering (Singer and Hunt, 1977). Large amounts of electricity, gas and water are wasted because of poor planning and management. Equipment is turned on first thing in the morning and much of it is left on all day. Large volumes of water are used, for example when defrosting frozen foods, cleaning vegetables, blanching vegetables. Of all forms of catering, research has shown that hotel kitchens are the most wasteful and

Appliance	Fuel	Power kW	Diversity % (1)	Date	Hours off	Hours on In use	Hours on Out of use	Daily energy consump kWh (2)
Steamer	Gas	20	50		3	2	3	50
Fryer	Elec.	15	50		5	2	3	22.5
							Total kWh	72.5

Notes:

(1) Diversity = $\dfrac{\text{average rate of energy use} \times 100}{\text{power rating}}$

A typical figure is 50% for cooking equipment and 20% for refrigeration. More accurate figures can be obtained from manufacturers.

(2) Energy consumption (kWh) = $\dfrac{\text{power rating (kW)} \times \text{diversity} \times \text{hours on}}{100}$

Figure 4.14   *An equipment audit*

may use two to three times more energy than other types of catering. Because of this there are excellent opportunities for savings without any negative impact on customers.

Where possible, localized energy meters should be installed in the kitchen in order to monitor consumption and to identify benefits from any changes made. Metering allows the kitchen energy consumption to be considered as a variable cost which can then be controlled (Unklesbay and Unklesbay, 1982). Another way of identifying the most significant energy users in the kitchen is to carry out an equipment audit, as shown in Figure 4.14. In order to do this, all energy-consuming equipment should be identified, and a note made of its power rating and the length of time it is in use each day (Fuller and Kirk, 1991, pp. 216–218). Ideally this would be set up on a spreadsheet to allow potential changes in energy consumption to be modelled.

Savings in general kitchen operations

Where there are several different kitchens throughout a hotel can these operations be centralized?

Is the cheapest energy source in use and has the most appropriate tariff been negotiated?

In the case of large electrical plant, is consumption affecting any peak demand charges?

Train staff to switch off equipment and lighting which is not needed in the next 10–15 minutes.

Match the capacity of equipment to the production needs since both over- and undersized appliances can lead to inefficiencies.

Fully load ovens, dishwashers etc. before use.

Several part-filled refrigerators are less efficient than one full one.

Train staff to match pan size to hob diameter since if a burner/element larger than the pan is used, energy is wasted.

Ensure that the bases of pans used on solid-top ranges have not become distorted.

Do not place hot food straight into cold rooms.

Frozen food should be defrosted in the refrigerator or cold room, which is good practice from a hygiene point of view and also reduces the energy demand on the refrigeration plant.

Maintain temperature of water from hot taps at 50–60°C.

Install flow restrictors on taps.

In areas like kitchens, extraction fans should be fitted with variable-speed motors. This allows staff to adjust the fan speed to the rate of extraction required. The correct use of extraction systems should form a part of staff training. If fans are left running when little cooking is taking place, the warm air extracted will need to be made up by the HVAC system. It is not uncommon, particularly at the start of the day, for staff to use catering equipment to supply background space heating to a kitchen which is too cold, partly because of excessive heat extraction from the kitchen. Because of this, energy is being wasted in two ways: the electrical supply to the extraction system and the wasted energy to the catering equipment. Sensible use of variable-speed extraction systems can reduce this waste.

Savings in food preparation

● Load and unload ovens, steamers and refrigerated cabinets as quickly as possible and do not leave doors open.
● Whenever possible, cover pots and pans while cooking.
● Ensure maintenance contract includes adjustment of gas burners.
● Install timers to switch off cooking processes automatically.
● Use internal thermometers to avoid the necessity to open the door when checking core food temperatures in ovens, etc.
● Segregate cooking equipment from refrigeration equipment within the kitchen.
● Encourage staff to use high-efficiency modern equipment where there is a choice; pressure cookers, combination ovens, brat pans etc.

Savings in sanitation

● Use hot water only when necessary.
● Do not use running water for cleaning purposes.
● Accumulate full loads for dishwasher, do not keep a flight washer running with no loads or small loads.
● Install heat recovery on dishwashers to recover energy from final rinse cycle.

Savings on refrigeration

● Turn off lights in cold rooms (decreases cooling load).
● Make sure all doors close properly, gaskets are in good condition.
● Keep coils free from ice build-up.
● Adjust defrost cycles to come on at night or other off-peak times.
● Transfer deliveries of chilled and frozen food into store as soon as possible to reduce any warming.

Energy recovery

The above measures are designed to minimize energy wastage as far as possible. However good these programmes of energy conservation through improved efficiency, some energy will always be wasted, usually in the form of heat. This heat will be wasted in association with the flue gases resulting from the combustion process, in hot air extracted from kitchens, laundries, sport and leisure centres and in waste hot water. It may be possible to recover a proportion of this waste heat. The recovered heat can then be recycled to provide part of the heating requirement for domestic water supplies or for heating systems. With all these systems, great care is required at the design stage if the potential benefits are to be obtained. The feasibility of recovering heat depends upon satisfying the following criteria (Fuller and Kirk, 1991):

1   The waste heat must be sufficient in quality (purity and temperature) and quantity.
2   Their must be a convenient use for that waste heat (not too far away and closely linked in time to the source).

3   The cost of recovery must be greater than the combined capital and running costs of its recovery.

The mechanics of recovery are reasonably straightforward. A heat exchanger is normally required to separate the waste heat from its source (combustion products, exhaust gases or waste water), to transport the heat to where it can be reused and to transfer the recovered heat to the chosen application (hot water supply, swimming pool water, warm air heating system, etc.).

There are a number of types of heat exchanger. Thermal wheels, which consist of a wheel made out of heat-absorbing mesh which rotates between inlet air and outlet air, can be used where the outlet air is reasonably clean and free from odours and grease and can recover up to 70 per cent of waste heat. They have been used in swimming pools to recover hot moist air which collects near to the roof of the pool and to recycle the heat into the make-up inlet air.

Where the air is contaminated with odours and or smoke, a system which separates the contaminated exhaust air from the clean air is essential. One way of doing this is to use an air-to-air heat exchanger (see Figure 4.15) where the two air flows are separated by metal plates which allow conduction of heat across the plate, while protecting the supply air from contamination in the waste air.

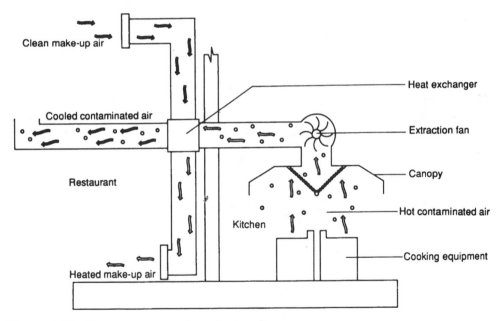

Figure 4.15   *Energy recovery using a heat exchanger*

An alternative to the air-to-air heat exchanger, particularly where the source of waste and location of re-use are some distance apart, is to use a run-around coil. In this type of heat exchanger, water (or some other heat transfer fluid) is pumped through a pipe going from the source of waste heat to the point where the heat can be re-used before returning to the starting point.

Where wasted heat is at a low temperature, such as the extract from canopies, room ventilators and swimming pools, this limits the usefulness of the recovered heat.

Although large volumes of heat may be extracted, it is often referred to as 'low-quality heat' because it is diluted with large volumes of air. To recover heat from a large volume of air which is at a low temperature is difficult and often not worth the cost involved. One of the problems with passive systems, such as run-around coils, is that the maximum temperature which can be delivered is that of the extracted air. Thus, if air from a canopy is at a temperature of 30°C, the maximum theoretical temperature which can be transferred to the water in a pre-heat tank is also 30°C.

If higher temperatures are desirable, then an active device, such as a heat pump is required. This works rather like a refrigerator, extracting heat from a source at a low temperature and delivering the heat to a region at a higher temperature. In this way, a heat pump can raise the temperature of waste heat, but in order to do this an additional form of energy, usually in the form of an electrical or gas-powered compressor, is required (see Figure 4.16). For example, air can be extracted from a canopy at a temperature of 30°C and be used to preheat domestic hot water supplies to a temperature of 50°C. Whether this is effective or not depends on two factors, the first of which is the ratio of energy recovered to energy which must be supplied to the heat pump. This ratio is known as the 'Coefficient of Performance', which is typically of the order of 2.5 to 3. A figure of 3 would mean that for every kilowatt-hour of electricity or gas supplied to the heat pump unit a output of 3 kWh of heat would be supplied to the heating or domestic hot water system. The second factor is the relationship between the quantity of heat wasted, together with the time at which this waste heat is generated and the time at which the recovered heat can be re-utilized. Heat is usually expensive to store for long periods of time.

As with any investment, the financial feasibility of an investment in heat-recovery measures will depend on the payback time for the capital installation based on fuel savings in supplying the space heating, pool heating or domestic hot water. As energy becomes increasingly expensive, this payback time will shorten and investment is likely to return a greater benefit in years to come.

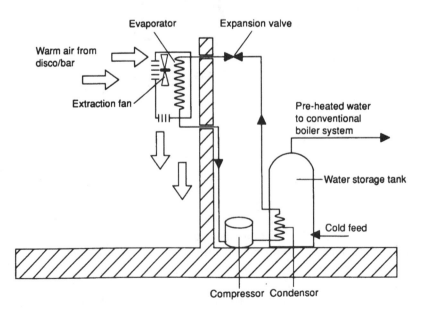

Figure 4.16   *Use of a heat pump in an energy-recovery system*

Case studies

Ramada International Hotels and Resorts

Ramada has energy-conservation initiatives running throughout the group. In the Caribbean, solar energy is used to power lighting and ceiling fans. In tropical areas sliding doors and windows are fitted with connectors so that when they are opened, the air conditioning automatically switches off. Timers are used to shut off power to lights and equipment when not in use, while in the kitchens all non-essential equipment is turned off between 2pm and 5pm.

In the UK, Ramada hotels conserve energy through the use of low-energy lighting products. The Ramada Hotel, Reading has replaced existing boiler burners with a more efficient burner, saving an average of £300 per month on gas – a 12-month payback. Thermostatic valves have been fitted to radiators in Manchester, and the air-conditioning uses waste heat in a run-around coil system. Corridor lights operate on switches, and outside lighting uses photo-electric cells. Energy consumption is logged throughout the group.

Inter-Continental Hotel group

In a pilot project the Inter-Continental group installed gas, electricity and water meters in individual departments in three of its hotels, to assess the potential for financial savings. Under this scheme departments were charged for their energy consumption and held accountable for improving performance. (Sub-meters had been installed in the majority of hotels in the course of previous energy and water audits.) Benchmarks were established and the amount of waste or inefficiency calculated for each department, enabling realistic targets to be set.

Hotel general managers were requested to continue with the installation of meters and the scheme was extended to 75 per cent of hotels. Initial technical problems were ironed out, and the benefits of the scheme became clear – of the $5 million saved (4 per cent each year for 3 consecutive years) in seventy-six hotels since the campaign started, approximately 60 per cent can be attributed to this approach. A wealth of technical data and feedback from the hotels has provided valuable information for setting future targets and priorities.

Forte plc

In 1977, Forte introduced a plan to reduce the amount of energy used by the group. The first phase of the plan was a staff awareness programme, designed to show how energy could be saved by good housekeeping practice. The next phase was to deal with items of maintenance that could be handled within the normal maintenance budgets. The third phase was to deal with all items requiring some capital investment, beginning with those items giving paybacks of 3 months or less, going on to 6 months and so on. The company estimates that its planned approach to energy conservation has saved a considerable amount of money – substantially reducing its yearly energy costs.

Forte has also experimented with the installation of combined heat and power machinery. In 1987 trials at the Manchester Post House gave energy cost savings of £7000 per annum. A major installation at the Heathrow Crest Hotel is expected to give a payback of 2.3 years. In one month alone the installation saved the hotel more than £2657.

Estimates suggest that this performance will reduce carbon dioxide emissions into the atmosphere by 115 254 kg annually.

Regency Inter-Continental, Bahrain

When replacing the hotel's electric steam boiler after 12 years of service, the Regency Inter-Continental in Bahrain decided to install two oil-fired boilers. This achieved a reduction in steam cost and other secondary benefits, such as using flash steam, which was previously wasted, for heating hot water calorifiers.

The new boilers have been in operation since May 1992. The following is an evaluation of the cost savings that have been made for one month:

1   Operational cost comparison for laundry steam:

Fuel oil consumption for May 1992	
18 812 1 × US$0.2112	= US$3988
Electric boiler – electrical consumption	
for same month last year	
158 900 kWh × US$0.042	= US$6673
Monthly saving	= US$2685

2   Flash steam supply to three hot water calorifiers:

The 8.5 bars high pressure and 6 bars medium pressure condensate returns from the laundry pass through two flash vessels (made in-house), where flash steam of 2 bars pressure is generated. This new source of energy, which was previously wasted, is now being used to supply three hot water calorifiers.

The boiler will only operate at 2 bars controlled pressure during the laundry off times. (An additional pressure controller was installed in each boiler.)

Fuel oil consumption for May 1992	
28 051 × US$0.212	= US$ 595
Electrical consumption for same month	
last year 54 740 kWh × US$0.042	= US$2299
Monthly savings	= US$1704

Another benefit derived from the new boilers is that of maintaining the steam pressure at 7.5–8.5 bars at all times, resulting in a reduced contact time on the flatwork ironer, thereby increasing the speed by 30 per cent with an improvement in laundry production in terms of quality and quantity.

L'Hotel, Toronto

L'Hotel, a Canadian Pacific hotel in Toronto, has developed a comprehensive programme to conserve energy. With the co-operation of Ontario Hydro (the regional electricity company) L'Hotel switched its 40W fluorescent tubes to 34W tubes, resulting in a pay-off in energy savings in the first year of $25 000 for an investment of $10 000 – in addition, the hotel qualified for a rebate frm Ontario Hydro for energy conservation ($1900). The installation of a state-of-the-art building management computer system has improved energy saving efforts by shutting down lighting and boilers at off-peak times, resulting in savings of $8300 on the gas bill.

Le Meridien, San Diego

At Le Meridien, San Diego, energy conservation is a key topic. An off-premises, independent co-generation system provides prime rate electricity to the hotel. Heat, which is a by-product of the co-generation engine, is used to run HVAC absorption chillers and produce hot water for both heating and sanitary use.

Classic light bulbs and fluorescent tubes have been replaced by compact low-density light bulbs, resulting in lower energy consumption and related costs while preserving quality lighting within the premises.

Hyde Park Inter-Continental, London

Since the appointment of the Regional Chief Engineer at the London Hyde Park Inter-Continental Hotel in 1981, changes to the hotel's plant have meant great financial savings for the group. Some of these changes include:

1 Recovery of flash steam from the laundry and leaking steam traps:

   Capital cost: nil   Annual savings: $10 000

2 Replacement of incandescent lamps in emergency stairs, corridors, etc. with low-wattage energy-efficient fluorescent lamps:

   Capital cost: $10 000   Annual savings: $10 000
   Simple payback: 1 year

3 Recovery of water previously lost by draining the kitchen cold storage rooms:

   Capital cost: $40 000   Annual savings: $20 000
   Simple payback: 2 years

4 Conversion of kitchen and extractor fans from fixed volume to variable-speed drives:

   Capital cost: $20 000   Annual savings: $12 000
   Simple payback: 1.7 years

5   Re-running of pipework to chillers so that a balanced flow could be obtained for condenser water (the pumps worked against each other):

Capital cost: $10 000   Annual savings: $5000
Simple payback: 2 years

6   Improving combustion control and operating methods for the two boilers leading to better heat transfer:

Capital cost: nil   Annual savings: $13 000

7   Installing a small air compressor for night operation:

Capital cost: $4000   Annual savings: $2000
Simple payback: 2 years

8   Complaints about the air-conditioning led to the replacement of the obsolete pneumatic control system, which was difficult to maintain. The cost would have been $145 000. Instead it was decided to install a state-of-the-art TREND DDC building automation system.

Capital cost: $326 000 (–145 000)   Annual savings: $51 000
Simple payback: 3 years

Some of the benefits of the system are:

● Air-conditioning temperatures in all public areas are now accurately controlled, whereas previously constant adjustments had to be made.
● Chillers and heaters, which used to be started manually and ran for many more hours than necessary, now start automatically as required.

Ramada Renaissance Hotel, Melaka

At the Ramada Renaissance Hotel, Melaka, fuel consumption was very high, due to poor heating of the laundry equipment and recovery of the condensate return. New pressure-reducing valves were installed in the laundry equipment, cutting fuel consumption, and new steam traps were also installed.

As a result of these measures, the consumption of other utilities, such as water and electricity, has been reduced. Fuel consumption was reduced from 339 774 litres at a cost of US$72 808 in 1989 to 261 400 litres costing US$56 028 in 1990, saving US$16 779.

Inter-Continental, Sydney

The Inter-Continental hotel in Sydney has discovered that in the laundry the same cleanliness can be achieved by washing at 60°C as at 96°C. This has resulted in an annual gas saving for the hotel of approximately A$24 000.

Sheraton Hotels, Hawaii

The Sheraton hotels in Hawaii have developed a bottom-line-oriented energy awareness programme and a recycling programme that aims to create both a supply and demand for recycled goods without an increase in net costs. Sheraton's four hotels represent 4500 rooms which translates into considerable costs in waste haulage and disposal. Since the recycling programme began, hotel waste tonnage has been reduced by an average of 13 per cent resulting in $7000 per month savings. In an arrangement with Hawaii Environmental Transfer, recyclable commodities are collected at a central loading dock. Each month an average of 14 tons of cardboard and 4 tons of office paper are collected.

The Sheraton Waikiki has reduced electricity costs by nearly $26 000 simply by replacing bulbs with fluorescent fixtures in the public areas. Additionally lobby areas and the loading docks have been targeted for lighting upgrade, with nearly $14 000 in yearly savings as an inducement.

References and further reading

Energy Efficiency Office (1990). *Energy Efficiency in Buildings: Hotels*, London: Department of Environment.

Energy Efficiency Office (1993). Energy Consumption Guide 36: *Energy Efficiency in Hotels*, London: HMSO.

Energy Efficiency Office (1994a). Good Practice Guide 136: *Is Your Energy Use Under Control?* London: HMSO.

Energy Efficiency Office (1994b). Good Practice Case Study 245: *Energy Efficiency in Hotels – Energy Efficient Space Heating and Hot Water*, London: HMSO.

Energy Efficiency Office (1994c). Good Practice Case Study 260: *Energy Efficiency in Hotels – Occupancy Linked Controls*, London: HMSO.

Fuller, J. and Kirk, D. (1991). *Kitchen Planning and Management*, Chaper 10, Oxford: Butterworth-Heinemann.

HMSO (1994). *Sustainable Development: the UK Strategy*, Chapter 19, London: HMSO.

IHEI (1993). *Environmental Management for Hotels*, Oxford: Butterworth-Heinemann.

Kirk D. (1987). Computer systems for energy management in hotels. *International Journal of Hospitality Management*, **6** (4), 237–242.

Kirk, D. and Milson, A. (1982). *Services, Heating and Equipment for Home Economists*, Chichester: Ellis Horwood.

Singer, D. D. and Hunt R. O. (1977). Energy use in the catering and food industry. In Glew, G. (ed.) *Catering Equipment and Systems Design*, London: Applied Science, pp. 85–94.

Unklesbay, N. and Unklesbay, K. (1982). *Energy Management in Food Service*, Chapters 15 and 16, Westport, CT: AVI.

Unklesbay, N. and Unklesbay, K. (1985). Energy waste: how large is the problem and how can it be reduced? In Glew, G. (ed.) *Advances in Catering Technology – 3*, London: Elsevier, pp. 167–180.

5 Management of the indoor environment

The significance of the indoor environment

In this chapter we will consider the indoor environment as including the quality of the air within the building, lighting levels and noise levels in relation to the comfort, wellbeing and health of the occupants of the building. These occupants include employees, guests and any other groups of people, such as outside contractors, who may be exposed to the environment. While some of this subject matter relates to occupational health, this is a much broader term covering many factors other than the environment, and is beyond the scope of this book.

There are two levels of responsibility both to those individuals working in the building and to those who are guests of the establishment, the first covering the essential requirements of health and safety and the second, the establishment of conditions which provide comfort. In terms of health and safety, the essential legal requirements for the UK are covered by The Health and Safety at Work Act 1974 and the Health and Safety at Work Regulations 1992. These are augmented by the Control of Substances Hazardous to Health Regulations 1988. Although those regulations control the use of all types of hazardous chemicals used in the workplace, they have a particular significance in relation to occupational lung diseases caused by exposure to dust, smoke and chemicals (HSE, 1994).

Comfort is a difficult term to define since conditions of comfort vary from one person to another and are also related to the activity of the person (Kirk and Milson, 1982). We can define comfort in objective terms such as air temperature, relative humidity, rates of ventilation, absence of impurities from the air, lighting levels and noise levels. However, in practice, individuals are tolerant of a range of values of any of these parameters (see Figure 5.1). If conditions fall within this range of values, the individual is unlikely to be aware of a feeling of comfort but, on the other hand, if the condition falls outside the range, then they are likely to be aware of discomfort.

For example, a person sitting in the hotel lobby reading a newspaper may be quite unaware of temperature changes between 18°C and 20°C but if the air was hotter or colder than this, they may feel discomfort. This is further complicated by the conditions of ventilation. A temperature of 20°C with poor ventilation might lead to a feeling of drowsiness and discomfort. The same temperature with adequate ventilation may be acceptable and therefore not noticed. A person who is undertaking

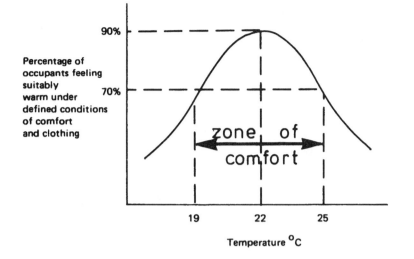

Figure 5.1 *A range of values for comfort*

a strenuous activity in the gym or leisure centre is likely to feel more comfortable with lower temperatures and higher rates of ventilation. This is because a person who is at rest will generate only 20–40 W/m2 of heat at the body surface, compared to 100 W/m^2 for someone taking gentle exercise. Comfort levels also depend upon the clothing worn. This makes the determination of optimum temperatures in lobby areas difficult, since some people will be wearing outdoor clothing while others may be in shirtsleeves.

Optimum comfort levels vary from one individual to another and are related to factors such as gender and age. In general, women prefer higher room temperatures than men and older people higher room temperatures than do younger people. In practice, all that we can do is to set temperatures and ventilation at average optimum levels for the type of guests we have and in relation to the activity taking place in that area of the hotel. It is also important to provide individual control over temperature and ventilation in the bedroom in order to allow the guests to set levels for themselves. Some people sleep best in a cool, well-ventilated room whereas others prefer a higher temperature. By giving the guests flexibility over the temperature and ventilation, this may avoid the need for them to open windows, which can defeat the most sophisticated BEMS and lead to a waste of energy.

Chemical hazards

A hazardous chemical is any substance that can cause injury, impairment to health or death to living organisms, or which may damage the environment. A number of hazardous materials may be used in a hotel's daily operation and their use may also mean the generation of hazardous waste. In view of the dangers associated with hazardous materials and their waste, it is important that they are handled, stored and disposed of carefully. Materials may be considered to be hazardous because they are:

- Toxic – a substance that can cause damage to health, physical or mental impairment or even death when inhaled, ingested or absorbed e.g. pesticides and herbicides
- Flammable – a substance that can be easily ignited by sparks or flames to cause fires, (e.g. solvents and fuels)
- Explosive – a substance which is capable, by chemical reaction within itself, of producing gas at such a temperature and pressure as to cause damage to the surroundings
- Corrosive – a substance which destroys other materials by chemical reaction, or causes burns to human tissue
- Infectious – a micro-organism which either causes an infection or produces toxins.

All materials, chemicals and substances that are used in the hotel and which may have a harmful effect on the environment should be identified, using a form similar to that shown in Figure 5.2. This can be done department by department either by hotel staff or by a contractor. Particular substances which must be identified include cleaning agents, pest and rodent control chemicals, paints, varnishes and lacquers, solvents, maintenance chemicals, chemicals from fire extinguishers, fertilizers, washing powders and sprays. Software packages are now available to generate and store data sheets in the form of a database, which makes updating easier. It should also be possible to obtain data sheets from the manufacturers of hazardous chemicals since, as of September 1993 in the UK, they are required to provide this information, under the Chemicals (Hazard Information and Packaging) Regulation (CHIP), which itself satisfies the requirements of the EC Substances and Preparations Directive.

In all cases the chemical content should be identified. Where materials are seen to be dangerous or harmful to the environment, safer alternatives should be found wherever possible. Where this is not possible, procedures should be established for the handling, storage, use and disposal of these materials.

In the UK there are strict controls governing the use of dangerous chemicals in the workplace, the Control of Substances Hazardous to Health Regulations (COSHH) 1988. It has been recognized that people who have regular contact with dangerous chemicals need safeguarding from the possible short- and long-term effects of these chemicals and that these risks may be immediate or may have delayed effects. The COSHH Regulations were introduced in order to safeguard people in working environments from any effects arising from the use of hazardous substances. The materials covered by this legislation are described in an earlier piece of legislation, the Classification, Packaging and Labelling of Dangerous Substances Regulations 1984.

Under the provision of the legislation, the employer must ensure that an assessment of all substances is carried out in order to identify two pieces of information, the potential risks from that chemical and any measures that need to be taken in the case of spillages or accidental exposure. Different risks may arise if the chemical is inhaled, ingested or comes into contact with the skin.

Following the assessment, the employer is required to ensure that exposure is either prevented or, if this is not possible, controlled. An important aspect of this control is that all employees who are likely to use, or be exposed to, hazardous chemicals must be trained about the risks and the necessary precautions to be taken when handling these chemicals, and the procedures to be followed in the case of spillage or exposure.

PRODUCT NAME:	PRODUCT CODE:	DATE OF ISSUE:

COMPOSITION:
Ingredient   Weight %
(list ingredients)

HAZARD IDENTIFICATION
(list hazards including flammability, harmful if inhaled, ingested or comes into contact with skin)

FIRST AID
Contact with skin
(indicate treatment)

Contact with eyes
(indicate treatment)

Ingestion
(indicate treatment)

Inhalation
(indicate treatment)

FIREFIGHTING MEASURES
(indicate type of extinguisher and other measures)

ACCIDENTAL RELEASE MEASURES
(indicate required action to control and treatment of spillage and release into atmosphere)

HANDLING
(instructions to handlers about avoiding contact, avoiding breathing fumes, not smoking, eating, etc.)

STORAGE
(indicate required storage – temperature, ventilation, away from food, etc.)

PERSONAL PROTECTION
(indicate need for protective clothing, respirator, etc.)

EXPOSURE LIMITS
(maximum concentrations for long-term and short-term exposure)

PHYSICAL AND CHEMICAL PROPERTIES
(data obtained from manufacturer on physical properties (viscosity, boiling point, flash point, viscosity, solubility in water) and stability and reactivity)

TOXICOLOGICAL DATA
(specific effects resulting from ingestion, contact with skin, eyes, inhalation – short-term and long-term exposure)

ECOLOGICAL INFORMATION AND DISPOSAL
(effect of chemical on waste water – biological oxygen demand, if biodegradable, methods of disposal – drains, incineration, special or legal requirements)

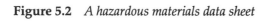

Figure 5.2 *A hazardous materials data sheet*

Also, the data sheets should be readily available to allow immediate action in case of accidental contact or spillage of the material.

Although many of the materials covered by the above legislation are not commonly found in the hotel environment, there are some cleaning materials and agents which do pose a significant risk to employees. These include oven cleaners, bleaches, detergents, drain cleaners, insecticide sprays and sterilizing agents together with chemicals used in the gardens and grounds such as pesticides, weedkillers and fertilizers. In the health club, chlorine-based disinfectants may be used. Engineering and maintenance departments are likely to use a variety of hazardous materials including solvents, acids, oils, grease, paints, wood preservative, hydraulic fluids, fuels and adhesives. Chemicals in the office areas of the hotel may include:

- Fumes from photocopier
- Typewriting correction fluid may be 1,1,1 trichloroethane, which is an ozone-depleting gas and can also cause breathing problems among staff
- Solvents, printing inks.

One group of chemicals of particular concern are the polychlorinated biphenyls (PCBs). They are found in a wide range of electrical equipment, such as transformers, capacitors, switches and voltage regulators, where they are used as insulators. They had been used for many years before it was discovered that they can cause a number of health-related problems. It has also been found that they are building up in the food chain because they are not easily biodegradable and that they cause cancer in animals. People exposed to the compound have complained of symptoms ranging from dizziness, nausea, eye irritation, bronchitis and digestive problems to more serious ones such as liver damage and chloracne, a painful and disfiguring skin infection. An inspection of all capacitors and transformers should be conducted by a qualified person and any found to contain PCBs should be replaced and the old equipment disposed of by a licensed agent.

Pesticides and herbicides are a group of chemicals used to kill unwanted life forms. They are widely used in hotels in kitchens, waste storage areas, guest rooms and hotel grounds. While some are harmless, most can cause a range of health problems in humans and animals including eye, lung, throat and skin irritation, dermatitis and poisoning. There are also long-term effects such as cancer and birth defects. Many of these chemicals are inert and build up in the food chain and in water supplies. To safeguard the welfare of guests and employees and to protect the general environment, where possible these dangerous chemicals should be phased out or replaced by less hazardous ones. Where this is not possible strict controls over the storage, use and disposal should be instigated. This requires an undertaking to:

- Identify which pesticides and herbicides are being used
- Determine whether their use complies with regulations
- Ensure that the mode of storage, use and disposal safeguards health and the environment
- Assess alternative methods
- Carry out specific checks and actions relating to storage, preparation, application and disposal.

Pesticides and insecticides are used to control:

  Mosquitoes
  Flies
  Cockroaches
  Ants
  Rodents
  Garden pests.

Alternatives to pesticides and herbicides include:

  Biological control through the introduction of predator species
  Cultural control by traditional good practice
  Physical control using measures such as traps and UV exterminators.

There is a conflict in this aspect of hotel management, because many environmental health officers are reluctant to sanction alternatives to the use of chemicals for the control of micro-organisms and pests in areas such as kitchens and stores. Where alternatives to chemicals cannot be found, proper management of the chemicals is essential. Manufacturers and suppliers should provide full details of the hazards associated with their products. They should also supply precise information relating to storage temperatures and conditions, first aid requirements, dilution factors for use and dosage rates.

Management should ensure that strict compliance with these rules is always undertaken. It is essential, for example, that all equipment used for the storage, dilution and application of the chemicals is unique to those purposes. Food and drink containers should never be used to store the chemicals. Treatment should be carried out only by competent and trained staff or contractors. Garden spraying work should be undertaken only when the climatic conditions are correct and when no hazard to life, other than the targeted species, is possible. For example, flowers should not be sprayed with insecticide when they are being visited by bees and insects. Protective clothing and respirators should be worn and all the equipment frequently checked for faults. Access to other than essential personnel during the time of treatment should be prohibited.

Air quality

The significance of indoor air quality

The quality of air inside a building is of great importance because of its effect upon the occupants of that building, be they employees or guests. The control of this aspect of the environment affects both our comfort and our wellbeing. We can measure indoor air quality in terms of the proportion of normal air gases and the concentration of pollutants.

The quality of air inside a building cannot be taken for granted. A variety of illnesses have been traced to indoor-air contaminants, where poorly maintained facilities have too often been implicated. Although health concerns in most facilities are not critical, poor indoor air quality can commonly lead to comfort complaints, decreased productivity and even poor health. The importance of indoor health quality can be gauged by the fact that, on average, we spend 90 per cent of our lives indoors while breathing some 10–20 m³ of air daily.

The quality of air inside a building is a combination of pollution from the air outside the building, brought into the building along with the make-up ventilation air, and the pollutants generated from sources or activities within the building. While external pollution can play a significant part in the eventual indoor air quality, this chapter deals specifically with internally generated components. A further consideration is how we dispose of these undesirable materials; this is covered in Chapter 6, which considers external emissions.

In recent years the increased concern regarding indoor air quality has come primarily from the widespread use of mechanical ventilation and air-conditioning in modern buildings, with limited direct ventilation through openable windows. Emphasis on energy conservation has also decreased ventilation.

Global concern over the past 20 years about environmental quality in general, and air quality in particular, has led also to concerns about indoor air quality. Numerous incidents of sick building syndrome and major occurrences, such as the outbreak of legionnaires' disease in a Philadelphia hotel, have focused attention on specific indoor air quality problems. Guidelines and standards are being discussed and adopted around the world.

Health effects associated with poor indoor air quality depend upon the specific pollutants and their concentration levels. Typical minor symptoms include headaches, mucosal irritation (eyes, nose and throat), or respiratory discomfort. Severe reactions can include nausea or asphyxiation and prolonged exposure can lead to various systemic effects of toxic poisoning or to cancer of the lung or other organs. In general, for hotel guests the main problem is not long-term exposure to poor indoor-air quality, but rather acute exposure which causes annoyance irritation, allergic reactions and other immediate illnesses. For hotel staff, long-term exposure can be a problem. Ventilation is required to:

- Control the concentration of moisture/humidity
- Dispose of surplus heat
- Remove micro-organisms
- Remove vapours, odours and smoke.

Sick building syndrome (SBS) was recognized as an issue of concern by the World Health Organization in the 1980s, although a number of experts dispute its existence. Although it has been known in old buildings, it is largely a phenomenon of post-war buildings. The World Health Organization estimates that as many as 30 per cent of new or remodelled buildings have unusually high rates of occupant health complaints, with real physical symptoms but without clearly identifiable causes. Although often temporary, SBS can also require long-term investigation and remedial action. It represents a range of problems, but often related to the common theme of poor ventilation or inadequate control over the indoor environment. Factors which have been blamed include dust in the air, fumes from photocopiers, chemical emissions from furnishing fabrics and building materials and emissions from flickering lights and VDU screens. SBS is usually identified when a number of occupants of the building all suffer from similar complaints, which are relieved when they leave the building for some time. Symptoms includes rashes, asthma, allergies, headaches and dizziness.

Potential sources of air pollutants

Many gases and vapours may be found in the environment. They may be toxic, flammable or explosive. Gases are those materials which exist in the gaseous state at room temperature and atmospheric pressure. They may be found as fuels (natural gas), in refrigeration and air conditioning equipment (CFCs) and in specialist equipment used by maintenance staff and contractors. These gases may leak into the environment through the normal wear and tear of pipes and valves or through faulty maintenance or operation. They include:

1 *Combustion products:* These may include gases (such as carbon monoxide, nitrogen oxides, sulphur dioxide or hydrocarbons) and suspended particulates from boilers, cooking stoves, vehicle engines in garages and other combustion sources.
2 *Chemical vapours:* These may come from cleaning solvents (including those used in dry cleaning), pesticides, paints and varnishes, photocopier emissions. Vapours often result from the evaporation of a volatile liquid as found in paints, solvents and dry-cleaning fluids. Many are toxic at high doses and others, while not highly toxic, may have a narcotic effect.
3 *Building materials:* Such materials may include toxic substances, such as formaldehyde in foam insulation, textile finishes, pressed wood, fibre glass or mineral fibres, plasticizers, etc.
4 *Tobacco-smoking products:* People are adversely affected by 'passive smoking' and there is now a legal requirement on an employer to control this. Building decorations and fittings are also degraded by smoke. For these reasons, smoking is now banned in most offices and many hotels have designated areas of the hotel as non-smoking, allowing guests the choice. Most restaurants have smoking and non-smoking areas, although the effectiveness of this depends upon the separation distance and the effectiveness of the ventilation systems. Smoking areas should always be sited nearest to the extract point of the system, and non-smoking areas nearest to the ventilation inlet. Some restaurants are implementing total no-smoking policies.
5 *Radon gas and radon products:* These are released by the soil on which the building is situated or by stone (especially granite), cement or brick building materials.
6 *Methane gas:* This may come from decomposition of landfill materials if the building is sited on or near a landfill used for municipal waste or from leaks in the gas distribution system.
7 *Water vapour:* In humid climates, high humidity causes occupant discomfort and mildew, with consequent discoloration and odours, and damage to materials. In climates requiring space heating, low humidity reduces heating energy efficiency and leads to sore throats and other irritation. The humidity should be controlled to between 40 per cent and 60 per cent.
8 *Odours:* Even at concentrations below those of health concern, pollutants can cause annoying odours. As well as the chemicals listed above, naturally arising odours from human activities (bathrooms, cooking, leisure centres) also contribute to poor indoor-air quality.

There are also other contaminants in the air, such as:

1 *Asbestos:* a specific category of material found in older buildings, where it was used

as insulation and as reinforcement in plaster, paint, bitumen, mastic, resins, plastic and cement. It was also used in ceiling and wall panels and in fire resistant coatings (see Table 5.1). The danger with asbestos is that when the fibres are released into the air and inhaled they induce asbestosis and cancers. Where asbestos is found in a building, it requires special attention where it is deteriorating or when disturbed by repair work. Where this is suspected, consultants should be brought in to inspect and report on the condition of the asbestos. Where in a dangerous state, the asbestos may be removed or encapsulated. The removal process requires total isolation of the affected part of the building until it has been declared free of the fibres.

Table 5.1   *Asbestos-containing materials in buildings*

Generic name	%	Dates of use of asbestos
Surfacing material		
Sprayed or trowelled on	1–95	1935–1970
Preformed thermal insulating materials		
Batts; blocks; pipe covering:		
85% magnesia	15	1926–1949
Calcium silicate	6–8	1949–1971
Textiles: cloth		
Blankets (fire)	100	1910–now
Felts	90–95	1920–now
Blue stripe	80	1920–now
Red stripe	90	1920–now
Green stripe	95	1920–now
Sheets	50–95	1920–now
Cord/rope/yarn	80–100	1920–now
Tubing	80–85	1920–now
Tape/strip	90	1920–now
Curtains (theatre safety; welding)	60–65	1945–now
Cementitious concrete-like products		
Extrusion panels	8	1965–1977
Corrugated	20–45	1930–now
Flat	40–50	1930–now
Flexible	30–50	1930–now
Flexible perforated	30–50	1930–now
Laminated (outer surface)		
Roof tiles	20–30	1930–now
Clapboard and shingles		
Clapboard	12–15	1944–1945
Siding shingles	12–14	?–now
Roofing shingles	20–32	?–now
Pipe	15–20	1935–now

Table 5.1   *continued*

Paper products		
High temperature	90	1935–now
Moderate temperature	35–70	1920–now
Indented	98	1935–now
Millboard	80–85	1925–now
Roofing felts		
Smooth surface	10–15	1910–now
Mineral surface	10–15	1910–now
Shingles	1	1971–now
Pipeline	10	1920–now
Asbestos-containing compounds		
Caulking putties	30	1930–now
Adhesive (cold applied)	5–25	1945–now
Joint compound roofing asphalt	5	?–now
Mastics	5–25	1920–now
Asphalt tile cement	13–25	1959–now
Roof putty	10–25	?–now
Plasters/stucco	2–10	?–now
Spackles	3–5	1930–1975
Sealants fire/water	50–55	1935–now
Cement, insulation	20–100	1900–1973
Cement, finishing	55	1920–1973
Cement, magnesia	15	1926–1950
Flooring tile and sheet goods		
Vinyl/asbestos tile	21	1950–now
Asbestos/asbestos tile	26–33	1920–now
Sheet goods/resilient	30	1950–now
Wall covering		
Vinyl wallpaper	6–8	?–now
Paints and coatings		
Roof-coating	4–7	1900–now
Airtight	15	1940–now

2   *Dust or particulate matter:* Introduced with outside air or from internal activities, these may also contain micro-organisms and can be irritants, particularly to people with allergies or respiratory weaknesses. They can damage equipment and decor and will increase cleaning requirements. Airborne dust is a common contaminant of air supplies. Dust enters the building through open windows, ventilators and on the clothes of people, and it can be associated with certain food supplies and as a result of building and maintenance work. While most dust is harmless, it can cause irritation and discomfort. Some dusts are harmful, particularly to sufferers of illnesses such as asthma and bronchitis.

3   *Airborne micro-organisms:* Such organisms as *Legionnella pneumophilia* are primarily associated with moisture in air-conditioning and ventilation systems (see Chapter 3). Droplet infection is an issue in inadequately ventilated and crowded places.

Improving indoor air quality

The objective of an indoor air quality programme is to safeguard the health and welfare of both guests and employees while on the hotel premises by adopting air quality objectives and standards, establishing procedures for dealing with specific indoor air quality problems, and carrying out routine maintenance procedures. Monitoring indoor air quality is a technical matter, and if the company does not have the necessary expertise and equipment in-house, it may be that an outside company or consultant is required.

The first stage in establishing an indoor air quality improvement programme is to carry out an initial screening of air quality to identify any major air quality problems. This can be based on comments from employees and guests and may also require sampling for routine air-quality indicators. Comments from staff and employees should be analysed to look for any patterns connecting locations in the hotel, time of day or time of year. Any repeated pattern may indicate a systematic problem concerning the operation of the building.

Making single measurements of air-quality indices can be misleading since the air quality will vary throughout the day and also be affected by seasonal factors. For this reason, 'snapshot' measurement often does not represent a valid picture of the indoor air quality level. For example, hotel areas affected by vehicle exhaust gases from the garage will have higher pollutant levels when garage traffic is greatest. Any measurements should take account of variations related to the time of day or year.

Initial screening should include the following:

- Carbon monoxide: a measure of ineffective combustion of fossil fuels
- Carbon dioxide: results from human metabolism and will build up in overcrowded or underventilated areas of the building
- Humidity: resulting from human activity, cooking, leisure centres together with over- and underventilated buildings
- Particulates or dust: nearby traffic or industry, building and maintenance work
- Ozone: associated with fluorescent lights and photocopiers
- *Legionella*: cooling towers and air-conditioning systems (see Chapter 3).

Where other problems are suspected, this can be extended to include chemicals such as nicotine, organics, formaldehyde and radon.

Further diagnostic studies may be required to identify sources and specific corrective action and the assistance of a technical consultant may be necessary. Where building work or renovation is taking place in the hotel, the contractors should be involved in the indoor air quality programme to ensure that any disruption to the air quality of guests and employees is minimized. Where building work may involve the disturbance or removal of asbestos, there are specific health and safety requirements.

Once the problems have been identified, there are three basic approaches to improving indoor air quality:

1 Eliminate or reduce the pollutant source, perhaps adjusting the time of use in which the pollutant is generated.
2 Filter or purify the air.
3 Ventilate or dilute pollutants.

Costs

The cost of an indoor air quality programme will depend upon the nature of the problem. In some cases there may be a small financial implication, such as increased maintenance costs because of the need to replace air filters more frequently or to use a higher grade of filter. Other maintenance costs may include the adjustment of air supplies to boilers and catering equipment to reduce carbon monoxide emissions. The cost of building work may be increased by the need to bring in specialist contractors to strip out asbestos before the building work can commence. At the other extreme, a major redesign of the HVAC system may involve high costs.

On the plus side, an efficient indoor air quality programme will reduce the costs of cleaning air diffusers, improve lifetimes of materials within the hotel and reduce potential complaints and claims against the hotel. Guest comfort and satisfaction, as well as employee productivity, will improve.

Evaluation

There should be a periodic evaluation built into the indoor air quality programme. This should include monitoring of complaints from guests and employees and how they are handled. There should be routine checks of indoor air quality indices through inspection and/or measurement to ensure continued compliance with standards. Customer and staff surveys should also be used to measure their opinions on the effectiveness of the indoor air quality programme.

Vehicles should be checked for emissions of CO, NOx and visible smoke. Use of pesticides in the gardens and grounds should be controlled and used only where absolutely essential. Staff using these chemicals should be trained in their use and safe disposal.

Heating and ventilation

Thermal comfort is usually associated with air temperature, relative humidity and air movement. Ventilation is required both to provide oxygen for breathing and combustion and to remove contaminants such as smoke, smells and carbon dioxide. In situations where natural ventilation does not provide an adequate exchange of stale air from within a room with fresh air, mechanical ventilation will be required. Mechanical ventilation may be provided to draw fresh air indoors (inlet), to remove stale air from indoors (extraction), or a combination of both inlet and extraction ventilation (see Figure 5.3). In a modern building it is likely that inlet and extraction will take place through a ducting system (often referred to as a plenum) providing heating ventilation and air-conditioning (Figure 5.4). For existing buildings, additional air-conditioning can be provided using a packaged unit designed to fit on an external window or wall

(Figure 5.5). In addition to general room ventilation, localized ventilation, in the form of a canopy, will be required for some catering and laundry equipment.

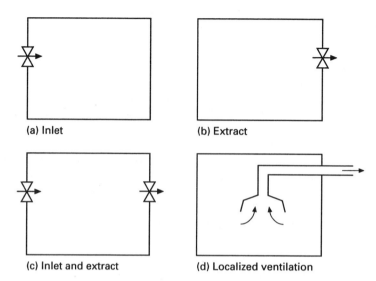

(a) Inlet   (b) Extract

(c) Inlet and extract   (d) Localized ventilation

Figure 5.3   *Types of ventilation*

The rate of ventilation, often measured in terms of air changes per hour, depends upon factors such as the number of people in the room, the nature of their activity and the potential contamination of the air caused by activities taking place in the room. This can vary from one air change per hour in bedrooms and public rooms up to twenty changes per hour in kitchens and laundries. Particularly where toxic materials are released (combustion products in a kitchen, dry-cleaning and other fluids in a laundry department), there is a requirement to ensure that the ventilation can remove these materials and that they are discharged in such a way as not to cause an additional hazard (see Chapter 6).

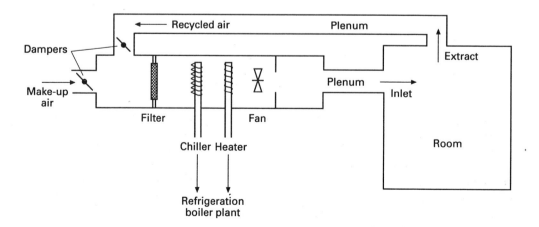

Figure 5.4   *A heating and air-conditioning ducted system*

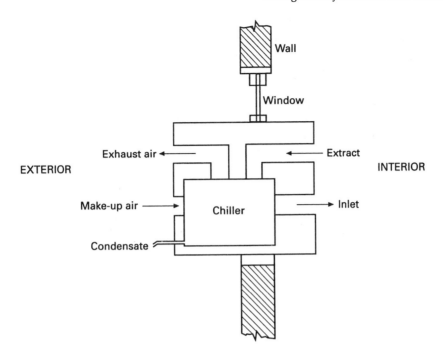

Figure 5.5   *A localized air-conditioning unit*

Noise

What is noise?

Noise is a result of the development of society and related activities in industrialization, population, traffic and other human activities. It is as much an environmental issue as pollution of water, air and soil. Over time, naturally present sound levels have been raised to increasingly higher levels to the point where it appears that we have reached an irreversible 'pollution level' of noise.

It is normal for people, particularly in cities and other artificial environments, to be constantly surrounded by noise. We can define noise as being any kind of sound that people consider undesirably disturbing, bothering or annoying and which can have a number of detrimental effects, including damage to health. This varies from noise within the workplace resulting from equipment and activities to noise in a leisure context from background music, traffic and the activities and noise of people themselves.

Noise can be spread from one part of a building to another through transmission of vibrations along building structures. The effect of noise can vary from a minor distraction at low levels to a severe health hazard for those exposed to very high levels or to lower levels but for a long period.

The health hazard resulting from noise is related to the intensity of the noise, its frequency and the duration of the exposure. The intensity of the noise is measured in terms of the decibel (dB) or sometimes pressure, in which case the unit is the Pascal (Pa). A level of 0 dB represents a sound threshold level, with a working area such as an office having a sound level of 40 dB and a jet engine 160 dB. When looking at measurements of sound intensity using the decibel it must be remembered that this is a logarithmic scale and therefore a small increase in the measured decibels can mean a large increase in the magnitude of the sound. For example, a sound intensity of 80 dB is ten times greater than one of 70 dB and a hundred times greater than one of 60 dB. A normal home environment may have a background level of 40 dB and an office 50-60 dB. In a city centre, the noise may rise to 70 dB and a passing heavy goods vehicle 90 dB at a distance of 15 m.

The frequency of a noise is measured in Hertz (Hz), where 1 Hz is the equivalent of one cycle per second. Higher-frequency sounds are expressed as kiloHertz (kHz) which is 1000 cycles per second. In practice, this frequency can vary from 20 Hz (the lowest frequency that can be detected by most people) to 18 kHz (the highest frequency which can be detected by most people). It should be stressed that people vary in their sensitivity to frequencies at the two extremes. Most noises are made up of mixtures of frequencies, and low frequencies introduce different types of health hazards from those at a high frequency. In terms of health and safety, sound levels for continuous and occasional exposure must be defined in relation to the type of noise hazard.

There are many sources of noise:

Traffic, e.g. road, rail, air
Construction
Industry and production
Other human activities including entertainment and sport.

In an industrialized or densely populated country we are constantly exposed to noise, which is present day and night at different levels. It affects people at work, at home, while travelling and while at leisure or on vacation. It must be realized that what is noise to some people may result from the leisure activity of others, such as the noise of a disco, a car engine or a motorcycle. Also, noise is relative: someone attempting to sleep in a hotel room may be much more aware of noise than if the same person were reading or dressing.

The effects of noise

In addition to being an irritant, noise can have physical and psychological effects on people. Physical effects depend upon the sound level and the duration of the exposure, but typical effects are:

Sound level: 65–120 dB
High blood pressure
Digestion problems, ulcers
Migraines
Depression
Sleeplessness
Neurasthenia
Circulatory disturbances
Irritability

Sound level: 85-120 dB
Hearing damage
Sound level: >130 dB
Direct, immediate damage to ear and deafness.

Particularly when considering employees, we must differentiate between constant exposure and intermittent exposure. All-day exposure to 90 dB may cause hearing impairment in a proportion of employees, whereas a level of over 100 dB may not cause damage if the exposure is for less that 15 minutes.

In addition to these physical effects on people, there can be psychological and physiological effects such as disturbed communications, distraction from concentration, the impediment of creative thinking, making people tired, creating or adding to stress, disturbing sleep, inducing bad moods and causing people to be aggressive and less productive. There can also be a reduction in the quality of life if an inadequate environment fails to facilitate recovery from stress at work. Also, when on vacation or in leisure time, noise can result in a significantly reduced level of enjoyment.

There are also financial consequences resulting from excess noise levels, since they can reduce the value of property, decrease employee productivity and efficiency and lead to a loss of business through repeat business and word-of-mouth recommendations. Noise from the hotel can also have a negative impact on the local community. This is discussed in Chapter 6.

Sensitivity to sound depends upon a number of factors:

Age
Sex
Mood
Present condition and health
Present level of stress
Time of day
Present activity
Acoustic factors
Understanding the necessity of the noise
Ability to control source
Expectation of quality of the environment
Attitude towards source
Education and training.

A programme for tackling noise

Quiet is the condition in which human beings generally feel well and in which they can relax, recover, rest or concentrate. Since the main objective of any hotel is to provide the best environment possible for its guests, a reasonably low sound level throughout the guest areas is extremely important. Equally important is the desired degree of privacy provided by a low level of sound transmission between adjoining rooms. Noise control will also improve employees' general wellbeing and productivity. Objectives of the action plan are to:

- Eliminate or minimize noise to create and maintain a suitable environment for guests and employees
- Prevent or minimize adverse psychological, physiological or physical effects to guests and employees
- Prevent annoyance of third parties (neighbours, tenants)
- Minimize possible revenue loss caused by annoyed guests who may decide not to return.

In a hotel which is being designed, care should be taken to physically separate, as far as is possible, noise-producing activities from noise-sensitive ones (Lawson, 1976). Where physical separation is not possible structures must be incorporated into the design which prevent the transmission of sound. Noise-producing areas include:

Kitchens
Laundries
Compressors, fans and mechanical plant
Delivery and refuse areas
Incinerators and boilers
Compactors
Ballrooms, function suites and discos
Bars and cocktail lounges
Lobbies
Public toilets
Swimming pools and leisure centres
Outdoor recreation areas

Areas sensitive to noise include:

Bedrooms
Meeting rooms
Conference halls.

In an existing hotel a noise audit should be the first step. This requires the identification of all possible sources of noise in the hotel, both interior and exterior, and the provision of a summary of known problems, partially based on an analysis of complaints. Noise can be tackled partly by changes in procedures, but investment in noise control may be necessary.

Measures to avoid noise will depend upon the specific problems identified, but could include the following:

- Determine day of week/time of day during which noisy work can be carried out
- Plan so that any known noisy activities coincide, leaving more quiet time
- Determine maximum sound levels within guest rooms for telephone ring, TV, radio and music sound levels and set accordingly
- Use telephone wake-up calls rather than alarm clocks
- Determine necessity for public paging and set restrictions for time of day and location
- Set schedules and maximum sound levels for musical entertainment in public areas for each outlet

Evaluate effects of noisy functions on guest room sound levels, especially when they take place at night

Consider relocation/elimination of disturbing night clubs and discos

Investigate causes of frequent false fire alarms and take remedial action

Check if better maintenance can reduce sound levels of elevators

Check that all doors are kept continuously closed, as appropriate

Insist that ear protectors are worn by employees/contractors involved in noisy work

Install time clocks for noisy ice machines on guest room floors, to switch them off at night.

Investing in noise control

When a sound wave strikes the surface of a material it may be reflected back, absorbed by the surface or transmitted through the material. This information can be used to control sound. We can contain sound by surrounding the source of the sound with reflective or absorbent material. Alternatively, we can build breaks into the structure of the building to prevent or damp transmission. Active measures are aimed at reducing the noise at source. These are the most effective and cheapest ways and may require new plant/major renovation/equipment replacement, such as:

Installing quieter motors and transmissions

Specially stiffened equipment structures

Damping

Low flow velocities

Well-designed ducts to prevent sound transmission from noisy areas to quiet areas through the ducting.

As sound can be carried through structures of the building, such as walls, ducting and pipework it may be necessary to change some of the structural aspects in order to introduce a reduction of sound transmission by structure, ducting and piping. Vibration can be a major source of noise transmission. Where possible, it should be controlled at source by, for example:

Encapsulation at source

Enclosure of either the equipment or the entrance to the room with highly sound-absorbing materials

Providing access to enclosures via easy-to-open and close hatches

Ensuring better construction of plant rooms, with noise-emitting sources isolated

Using sound-absorbent inner walls (mineral wool, fibre-glass, rubber)

Mounting noise attenuators on any cooling air opening

Isolating equipment in machines, through rubber mountings

Providing reinforced foundations for heavy equipment

Using special damping material, such as elastic panel mounting.

Passive measures can also be effective. These are aimed at protecting the recipient's ear. This can be done by the erection of sound barrier constructions which can reduce the transmission of sound. These measures must be taken whenever the emission source cannot be controlled and is at a level which cannot be influenced (e.g. aircraft, traffic and train). They should also be used where noise created in the hotel travels to undesirable areas.

The major decisions that determine the quality and quantity of noise in the hotel are made at the time of construction. Four factors play a major role:

1   Location and position of the hotel (influenced by owner)
2· Location of major noise sources in relation to other spaces requiring quiet (influenced by the architect)
3   Quality and design of construction, such as materials, routing of pipes and ducts, isolation techniques used (influenced by architect and engineers)
4   Quality of workmanship, such as sealing around pipes and ducting (influenced by the contractor)

However, considerable improvements can be made afterwards through:

● All measures listed above
● Better windows (also have benefit of improving thermal insulation)
● Closing all openings found in walls, ceilings or floor
● Installing gaskets, drop seals and automatic door closers on guest room door
● Replacing noisy fan coils by efficient quiet types
● Mini-bars to be supplied with absorber refrigerators rather than compression
● Replacing toilet flush valves with quiet flush tank
● Providing sound-absorbing barrier underneath bathtub
● Checking if sockets in adjoining rooms are offset
● Providing a quiet hairdryer.

Light

Lighting is normally provided by a mixture of artificial lighting and daylight. The intensity of light which falls on a surface is expressed in terms of the 'illuminance', which is a measure of density of light on an area of a surface. The unit of illuminance is the 'lux' or lumen/m². For staff carrying out detailed work, illuminance values of between 500 and 1000 lux are recommended, but this reduces to 200–300 lux for non-detailed work. Other areas, such as corridors and public areas, require between 100 and 250 lux. In bedrooms, the important factor is to provide ample localized control over lighting, allowing the guest full control.

In addition to its intensity, light is also categorized by its wavelength. Different forms of artificial light have different spectral distributions or 'colours'. The colour of lighting can be important in food preparation and service areas, since some types of artificial lighting can distort the colour of foods. The colour of light is often measured in terms of the colour temperature as in Table 5.2 (Kirk and Milson, 1982). As the colour temperature increases, the colour of the light changes from orange/yellow to white. In the case of fluorescent lighting, the situation is more complex because the light produced has a more complex distribution of wavelengths. For this reason, it is important to test types of fluorescent lighting with furnishings, crockery and food items to check for any undesirable colour distortion.

Table 5.2 *Colour temperatures associated with light sources*

Natural light	
Afternoon sunlight	4000 K
Noon sunlight	5000 K
Cloudy sky	6500 K
Blue sky	10000 K
Fluorescent	
White	3400 K
Plus white	3600 K
Warm white	3000 K
Daylight	4300 K

Other important factors are the direction of the lighting and whether it is focused, such as from a spotlight, or diffused. This is particularly important in lounge areas where a variety of activities will take place including relaxation, reading and holding conversations.

Effective levels of lighting are important because they:

● Add to the comfort experienced by the guest
● Improve the efficiency and effectiveness of employees
● Decrease the risk of accidents to guests and employees.

An excess of artificial lighting, or the absence of effective localized control over lighting levels, can lead to excessive energy consumption.

Non-ionizing radiation

Most forms of radiation form a health hazard but, in terms of hotel operations, there are only a limited number of sources of radiation, typically from microwave ovens (microwave radiation), lasers in printers (visible radiation) and sun beds (ultra-violet radiation). Ionizing radiation, usually associated with radioactivity, is not commonly found in the hotel industry, other than emissions of radon from certain types of building materials.

Microwaves are a type of emission known as non-ionizing radiation (Ashton and Gill, 1992). With non-ionizing radiation, the harmful effects are thought to be due mainly to the thermal effects resulting from the fact that the radiation causes a heating effect. The eyes are particularly sensitive and exposure can result in a symptom similar to a cataract. Routine measurements of radiation leakage from microwave ovens should be included in maintenance contracts. Most leakage occurs around oven door seals, particularly if spillages of food are allowed to accumulate. The recommended maximum exposure level is 100 W/m^2.

Lasers can cause damage to the retina of the eye as well as burning of the skin. Most commercial sound equipment and office printers which use lasers have sufficient protection for the normal users, but staff must be trained not to carry out anything other than routine maintenance as described in the manufacturer's handbook. Any repairs should be undertaken by individuals who have received specialized training.

Ultra-violet radiation can cause burning of the skin, skin cancer and damage to the eyes. There are three categories of UV radiation, based on the wavelength of radiation:

UV A 400–315 nm
UV B 315–280 nm
UV C 280–100 nm.

In relation to sun beds, staff operating this equipment should be given strict training covering the hazards associated with, and operational controls related to, specific pieces of equipment.

Case studies

The Hayman Great Barrier Reef Resort

The Hayman Great Barrier Reef Resort has its own power station, which has been designed to minimize pollution and noise emission: exhaust outlets are fitted with mufflers and the entire power station is soundproofed. In addition, computer equipment ensures minimum fuel usage.

The Hilton International Group

The Hilton International group is gradually eliminating high-tension transformers, which contaminate the cooling fluid with PCBs, and replacing them with dry-cast resin transformers. Hilton International recognizes that this is a very expensive process and that disposal of the transformers is problematic, but the group hopes to have replaced all the old transformers within the next 3 years.

Ramada International

Ramada International has begun to replace chlorine bleach in its swimming pools with a non-toxic ionization process.

References and further reading

Ashton, I. and Gill F. S. (1992). *Monitoring for Health Hazards at Work*, 2nd. edition, Chapter 7, Oxford: Blackwell Scientific.
Centre for Compliance Information (1978). *Noise Control in the Workplace*, Maryland: Aspen Systems.
HSE (1994). *A Recipe for Staff Health and Safety in the Food Industry*, Health and Safety Executive.

Kirk, D. and Milson, A. (1982). *Services, Heating and Equipment for Home Economists*, Chapters 7 and 12, Chichester: Ellis Horwood.

Lawson, F. (1976). *Hotels, Motels and Condominiums: Design, Planning and Maintenance*, Chapter 10, London: Architectural Press.

6 Materials and waste management

The need for materials and waste management

Despite a trend towards recycling in the last few years, an increasingly consumer-orientated society continues to generate great quantities of waste. This waste, and its disposal, has an ever-increasing effect on our lives, threatening our health, and the quality of our environment and placing a growing burden on business and national economies. As a result, the management and minimization of waste have moved high up the agenda for any commercial operation.

In particular, as landfill sites are being used up, they are becoming increasingly expensive and will become a huge operating cost in years to come. A waste management programme will help to reduce the amount of waste produced and, at the same time, save materials, resources, energy and money. The ability to mount a successful waste recycling scheme may depend upon factors such as local community programmes, together with local codes on waste disposal, landfill and the presence of incineration-based district heating schemes. As an example, some local authorities may have a scheme for the collection of Christmas trees at the end of the festive season, for shredding and conversion into forest bark or for incineration. It is important to be aware of local schemes and to cooperate with local initiatives.

It is now recognized that waste management is not simply a matter of disposing of unwanted and sometimes hazardous output and it is important that we differentiate between waste minimization and waste disposal management (Cummings, 1992). Waste minimization (i.e. making sure that the minimum amount of resources are wasted) must be given the highest priority because not only does it conserve resources, it also has the potential to save the hotel considerable amounts of money. Only when the volume of waste has been reduced to its absolute minimum should we then go on to consider those issues related to the responsible disposal of this waste.

Defined in these terms, waste management is a process that affects all stages of an operation, starting with product and process design and incorporating purchasing policy, stock control and production planning. A sensible waste management policy seeks to reduce the cost to the company of waste by maximizing the value of all resources. By maximizing the value of these resources and, in this way, reducing waste to the absolute minimum, the policy can give the maximum financial return to the company while, at the same time, reducing the need for landfill sites. This may be represented in the form of a waste hierarchy, as shown in Figure 6.1.

Figure 6.1   *A waste management hierarchy*

Waste management is the responsibility of all those employed in the organization, from those who design the products and services through all levels of operational management. By reducing waste, through accurate purchasing which is closely linked to production plans, some waste can be eliminated before it occurs, saving resources and their full purchase value and with the elimination of the costs of unnecessary procurement, transport, storage and processing. Production plans, which are related to accurate sales forecasts, can save both the cost of raw materials and the cost of processing the unwanted materials.

The above measures provide the best possible financial return but, however good these measures, some waste will still occur. Where possible, these waste materials should be re-used within the operation, even if at a lower value. Materials which cannot be re-used within the operation should, where possible, be collected for recycling. Materials that have an inherent energy value, but for which there is no market for recycling, may be used as part of a local incineration/power generation or district heating scheme. Finally, if none of the above are possible, it may be necessary to consign waste materials to landfill sites. This is the least preferred alternative since there is the double environmental impact in that the value of the material is lost and land is destroyed in the form of a landfill site. Whatever method is chosen for the disposal of waste materials, the hotel company has a duty to ensure that all waste materials are disposed of in ways which are legal, under requirements known as 'Duty of Care'.

The waste audit

Waste management programmes should commence with a waste audit, in which the quantity and type of waste produced in the hotel are assessed. A form, such as that shown in Figure 6.2, might be used to record all purchases and their potential for re-

Hotel product used	Reduce	Re-use	Recycle	Replace
Aerosols				
Air conditioner				
Aluminium				
Appliances				
Batteries				
Bedding				
Bleach				
Bleach bottles				
Brochures				
Cans (aluminium)				
Cans (tin)				
Carpet remnants				
Cleanser				
Clothing				
Computer paper				
Cooking oil				
Corrugated boxes				
Detergent boxes				
Dishes				
Disposable diapers				
Disposable pens				
Dry cleaner				
Facial tissues				
Fertilizers				
Fine paper				
Food (meat waste)				
Food (non meat waste)				
Food-packaging				
Furniture				
Glass bottles				
Laundry bags				
Leaves/Grass clippings				
Light bulbs				
Magazines and books				
Matches				
Mattresses				
Menus				
Metal				
Mirrors				
Motor oil				
Newspapers				
Office equipment				
Oil (kitchen)				
Oven cleaner				
Packaging				
Paints and solvents				
Paper cups				
Paper towels				
Pencils				
Pens				
Pesticides/herbicides				
Plastic bags				
Plastic bottles				
Plastic buckets				
Plastic shower curtains				
Pots and pans				
Printed matter				
Rooming booklets				
Shampoo				
Shoe bags				
Stationery				
Sterno				
Styrofoam				
Tissue paper				
Toxics				
Windowed envelopes				
Wood				
Xerox paper				

Figure 6.2   *A hotel product checklist*

use and recovery. Once this is known, the possibilities for a waste management programme can be assessed by considering each kind of waste and deciding whether it is possible to:

1 Reduce the quantity of waste produced, by avoiding overpurchase, defective purchases (zero defects) poor stock management and faulty production.
2 Eliminate waste resulting from overpackaged goods and by encouraging suppliers to change production processes to generate fewer unusable by-products.
3 Re-use items in their original form for the same or for a different purpose. Refillable bottles, cloth towels and laundry bags, washable napkins and rechargeable batteries can all be used many times as opposed to their disposable equivalents. Some materials may not be re-usable directly, but may be acceptable for re-use by other groups. For example, partially used shampoos removed from bedrooms may be offered to local charities.
4 Recycle: many waste materials can be extracted to meet a market demand. Recycling possibilities should be guaranteed by manufacturers and suppliers.

Product purchasing

Environmental considerations of purchasing policy

It is now recognized that purchasing policies can make an important contribution to environmental management. While traditional purchasing policy may concentrate on concepts such as suitability for use and value for money, current developments, often instigated as part of quality management programmes, concentrate much more on developing partnerships with suppliers to ensure that materials are of an appropriate quality and suitably packaged. The application of 'just-in-time' techniques recognizes the linkage between purchasing, quality products and waste. If these factors are taken into account, it is possible to build in sound policies which include factors such as sustainability of supplies, the use of recycled materials where appropriate and the minimization of packaging.

It is important to recognize the importance of links between input factors, such as purchasing decisions and output factors such as recycling and waste disposal. As indicated in Chapter 1, sub-systems are linked in a complex way and the nature of these relationships must be explored. This means that, at the stage of making purchasing decisions, we must consider the extent to which waste materials such as packaging can be reduced to a minimum and be recycled. For that waste which cannot be recycled, it is important to know how it is to be disposed of and to be certain that the waste does not generate toxic decomposition products. It is possible to consider re-useable packaging, such as buying light bulbs in tea chests and foam layers rather than in cardboard sleeves. This can provide savings both to the bulb manufacturer and to the operator.

Purchasing principles:

- Buy only what is really needed
- Buy material of an appropriate quality to reduce defective material

- Buy locally where possible to support the local community and to reduce transportation
- Buy in appropriate quantities – too small quantities will increase transport costs, too large may result in spoilage and, in any case, increases stock-holding costs
- Buy for energy efficiency
- Buy recycled products or products in recycled packaging where at all possible
- Buy products which are made from and packaged in materials that are recyclable
- Consider renting instead of buying
- Beware of exaggerated claims by vendors
- Favour products that are gentle on the environment, such as biodegradable detergents
- Avoid disposable products where possible
- Minimize packaging.

With these principles in mind, it is possible to audit all products purchased and to assess their environmental impact. Similarly, all contracts for services should be evaluated against this checklist. Any new contracts for products and services should include environmental impact as part of the specification.

Where harmful products are identified, alternatives should be investigated through suppliers and distributors. These alternatives should be evaluated in terms of both their intrinsic properties (such as price, reliability of supply, quality of product/ service) and their environmental impact.

Some practical steps to environmental purchasing

Some examples that can be incorporated into normal purchasing decisions include:

1 Wherever possible, all wooden items should come from a managed forest and the source of hardwoods in particular should be investigated with their supplier.
2 Imported foods should give an appropriate return to the community where they are grown.
3 All new equipment which requires an input of energy should have the maximum energy efficiency.
4 The capacity of any appliance purchased should be linked to the normal load of the appliance when in use, since appliances that are either over- or undersized will operate with a reduced energy efficiency.
5 Where possible and practicable, products should be purchased which contain the maximum amount of recycled materials.
6 As far as possible, identify products that have packaging materials that can be recycled.
7 In an activity where many batteries are used, consider the purchase of a battery charger and rechargeable batteries.
8 No equipment or materials that either contain CFCs or which use CFCs in their manufacture should be purchased.
9 Where fossil fuels are purchased (such as fuel oil, coal, gas), these should have a low sulphur content.
10 Any vehicles purchased should use lead-free petrol.
11 Petrol burning vehicles fitted with a catalytic converter are a better alternative than diesel engines.

The most difficult decision will be in those situations where the most environmentally sound product or service is not the best in terms of price, quality and reliability. As an example, if the effectiveness of a biodegradable detergent is poorer than the non-degradable alternative, purchase decisions may be complex. Clearly, a business cannot sacrifice large amounts of profit or product performance unless there are very clear business advantages which arise from use of products that have a lesser environmental impact. However, discussions with the supplier of the products/services may lead to improved performance of the environmentally sound product.

Operations management

As described above, the highest priority should be given to preventing waste, through sound design and planning before it occurs. The waste audit should provide an indication of the types of waste being produced and the areas of the operation where they are being generated. The processes leading up to this waste should be thoroughly investigated and changes to processing methods adjusted to minimize this waste.

Most forms of waste are evidenced by waste material which needs to be disposed. Just as important is waste that is not visible. Unfortunately, this will often not show up in a waste audit which looks at waste in the form of outputs. For example, an excessive use of detergents in a dishwashing machine or cleaning materials by the housekeeping department may not result in the generation of visible waste, but the process is equally wasteful. The excess of detergent not only has to be paid for but also ends up in waste water supplies where it may increase the cost of water treatment or result in higher biological oxygen demand and damage aquatic life. Similarly, waste food materials may be disposed of, not through waste bins but flushed down the drains using a waste disposal unit. This represents a loss of food and money together with an increase in the pollution of waste water.

In order to identify some of these 'invisible' sources of waste, it may be necessary to look at the relationships between usage rates and business activity using an input–output analysis, together with measures of efficiency. This is also an area where a comparison of consumption figures between the hotels in a chain can also throw light on areas of waste. For hotels which are not part of a chain, industry performance standards may be available from professional bodies and manufacturers' organisations.

One aspect of hotel operation which is known to generate waste is the production and service of food, where the hotel industry generates higher levels of food waste compared to many other types of catering. Research conducted in the early 1980s on food waste in the UK hospitality industry (Banks and Collison, 1981) indicated that 15.5 per cent of edible food was found to be wasted in restaurants and hotels, compared with an average of 11.4 per cent for the catering industry as a whole. This represents a high monetary value because of high value-added associated with the waste of prepared food (Collison and Colwill, 1986). It is also a waste of energy because the food has been transported, stored and cooked.

An investigation of the level of food waste should indicate the main causes of this waste. Is food discarded because it is beyond its 'sell-by' date or it is stale? This would indicate poor purchasing decisions or purchasing which is not closely related to production schedules. An alternative cause of this waste might be poor storekeeping and lack of proper stock rotation. All food stores should operate using the principle of 'first-in, first-out'. If the conversion of raw materials into finished dishes of food is poor, this may indicate wasteful preparation. Food which has been prepared and cooked but never consumed by the guest is perhaps the most wasteful, since not only is the food wasted, but so is the staff time in its preparation and the energy which was used to cook it. Areas like the buffet should be checked carefully to identify the level of waste. With tighter food hygiene regulations, there may be a tendency to discard food which has been on display too long. It would be much better to produce less food or smaller batches in order to match supply and demand better. Are guests leaving a lot of food on their plates? If so, this could indicate that portion sizes are too large, always assuming that the food is not being left because it is unacceptable to the guest. Whatever the cause, this should be investigated together with possible solutions to this problem.

Hotels, like most modern businesses, consume large volumes of paper. While recycling paper may seem the obvious first solution to this problem, in reality it is much better to investigate ways of either reducing consumption or re-using the waste. It may be possible to reduce the total volume of paper through an analysis of purchases and waste, using a form shown in Figure 6.3. For example, all management reports should be evaluated in terms of their length, the size of the circulation lists and the quality of paper used. Another good policy is to give all departments budgets for paper in order to manage consumption. Where possible, departments should be provided with double-sided photocopiers. Electronic point of sale (EPOS) systems in bars and restaurants may reduce the volume of paper required. Many companies are now attempting to make better use of electronic mail systems in order to improve communications but this also has the effect of reducing paper consumption. At a less sophisticated level of technology, chalkboards or whiteboards may be used for staff notices. However, the 'paperless office' may be some way off for most of us. In the meantime, it is important to evaluate all paper used in the business and see if lower-grade or recycled paper could be substituted for virgin paper.

Where waste seems inevitable because of the nature of the process, consideration should be given to the re-use of that material. For example, if a carvery leads to a waste of prime quality meat, ways in which that meat can be re-used should be investigated. Where possible, this re-use should maximize the value of the material. For example, it is better to re-use cold roast meats in 'quality' products such as sandwiches rather than in low-value dishes.

In some situations it may not be possible both to re-use materials and to maintain their value. Hotels which provide guests with soaps, shampoos and other detergents in individual packages, sachets or containers have no use for the remains of partially used containers. It may be possible to recover some of the value of the material, if not the cost associated with it, by donating these partially used containers to local charities. An alternative, which some hotel companies have used, is to fit bathrooms with bulk soap and shampoo dispensers. This means that guests can use as much of the material as they like but there is no waste. It also has benefit in that guests are not tempted to take home unused sachets or bottles. Old furniture which is no longer required by the hotel, particularly if it contains quantities of hardwoods, might have value to second-hand office and hotel furniture companies. If this is not possible, some charities have furniture shops which would find a use for the unwanted material.

Generation site	Type of waste paper	Estimated quantity	Temporary storage
Accounting			
Business centre			
Computer			
Engineering			
Executive office			
Front office			
Guests rooms			
Housekeeping			
Kitchen			
Mail room			
MIS			
Photocopy room			
Sales office			
Stewarding			

CONSOLIDATION

Computer paper   A _____

White ledger   B _____

Coloured ledger   C _____

Office mixed   D _____

Mixed paper   E _____

    TOTAL   _____

Types of paper
Computer paper: 18" wide.
White ledger: white bond, photocopy paper, laser paper, deposit slips, letter-sized computer paper.
Coloured ledger: coloured bond, photocopy paper, cheques, carbonless forms.
Office mixed: ledger, onion skin, envelopes, manila file folders.
Mixed paper: office mixed, magazines, newspapers.

Figure 6.3   *A waste paper audit record*

Environmental pollution

The importance of the control of environmental pollution

As we saw in Chapter 1, any real-world system interacts with the environment through a boundary which controls the flow of materials, people and information. For the environmentally aware company it is particularly important to control the flow out of the system of materials which may cause pollution of the environment. This pollution may vary in the level of severity, from causing a local nuisance because of noise, cooking smells and additional traffic, through to the discharge of environmentally damaging effluents (see Figure 6.4). As can be seen from Figure 6.4, some of the discharges and emissions are not damaging in a global sense, nor are they hazardous to the local community. In this category would come aspects such as the discharge of odours or noise pollution from generators or the disco. However, they are factors which have a large impact on the local community and therefore impinge on the image of the company, which in turn may affect the amount of business generated locally. Other aspects may cause damage to the local and/or global environment, such as the discharge of greenhouse gases, the contribution of traffic fumes to smog and the release of heavily contaminated waste water.

An assessment of all these overt and covert discharges should be undertaken in order to assess their impact. In terms of impacts which cause a nuisance to the local community, it is worth interviewing close neighbours of the hotel in order to gain additional information.

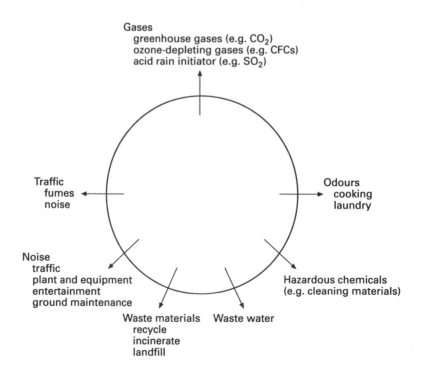

Figure 6.4   *Harmful and hazardous discharges from a hotel*

By following through many of the steps given in earlier chapters, many of these discharges to the environment will have been controlled, or at least minimized. Others may require some specific action on the part of the hotel management.

Waste water

Local regulations will determine the necessary segregation of waste water into categories such as foul (from WCs), waste (from sinks and baths) and storm or surface water. These local requirements will vary from one part of the world to another, as will the sophistication of water treatment. In some remote areas, waste water may be discharged straight into rivers. Care should be taken to ensure that waste water leaving the hotel is not unnecessarily contaminated. Periodic checks should be made to ascertain that spilt or unwanted fuel oils, paints, garden chemicals and detergents are not disposed of down the drains. In the laundry and housekeeping departments is it possible to use biodegradable detergents without any obvious loss of performance? The replacement of chlorine bleaches with perborate-based bleaches reduces the damage to aquatic life and may reduce the load on water treatment plants. The Royal Orchid Sheraton Hotel and the Towers Hotel in Bangkok have installed their own waste water treatment plant in order to avoid the need to pump untreated waste into the river.

Sewage treatment plants remove primarily suspended and organic matter. To a degree, they can also remove a number of undesirable substances. However, their ability to do so is very limited and costly. In addition, these plants differ greatly in their number and quality. Requirements for treating effluents have risen sharply in the last decades, as has the cost. In some cases sewer charges per unit already exceed fresh water costs.

Most countries are now accepting the principle of 'the polluter pays' and this trend is likely to continue. Therefore the elimination of problematic materials at the source should be aimed at whenever possible.

Gases, vapours and odours

The major gaseous emissions from a hotel are:

- Combustion products from the water boiler and HVAC systems
- Discharges from exhaust canopies in the kitchen(s) and laundry
- Discharges from extractor fans in public areas.

Two factors should emerge from an investigation of these emissions. In the first instance, these discharges should be minimized by factors such as improving boiler efficiency (see Chapter 4). Second, the nuisance value of these discharges should be minimized by locating chimneys or discharge vents as far away as possible from the neighbours of the hotel. Where emissions, such as cooking odours, cause a problem to the local community, it may be necessary to use absorbent filters in order to reduce the smell of these discharges.

Clearly, the hotel has a legal responsibility to ensure that the discharge of any harmful gases and vapours such as carbon monoxide and some cleaning solvents (see

Chapter 5) is done in such a way that they do not constitute a hazard. In this case, the requirements are exactly the same as those appertaining to employees under the COSHH Regulations (see Chapter 5). Another area where the hotel needs to take care is in the disposal of equipment (such as old refrigerators) which contain CFCs. If the equipment is going for scrap, the hotel management should ensure that the company carrying out the scrapping has the facilities for the collection of the CFC gases.

Noise pollution

The effects of noise pollution on the occupants of the hotel were discussed in Chapter 5. Many similar factors apply to the impact of noise from the hotel on the local community. Noise pollution can be a significant factor, particularly late at night. Noise arises from traffic entering and leaving the hotel, from plant and equipment (such as compressors, extractors and garden equipment) and from the activities of people in the hotel (such as leisure clubs, live entertainment and discos). Where noise pollution is a problem (and in extreme cases this may infringe local by-laws and result in enforced action), some of the sound control measures discussed in Chapter 5 may have to be considered.

Stored fuel and the environment

Oil and gas and many fuel products that derive from them are key resources in industry, transportation and energy production. There are a number of environmental hazards associated with these fuels. For obvious reasons, they are highly inflammable. As liquids they can easily cause fires and as vapours they fan out, generating an explosion risk. When fuel oil is released into the ground it coats soil with which it comes into contact. Plants are killed or damaged by contamination with fuel oil because of its toxicity and because it may trap plant nutrients in the soil. Fuel oil also seeps down into the water table and, because it is lighter than water, it sits on top of the groundwater and can move with it over great distances. Groundwater is often a source of domestic water supply and, where it has come into contact with fuel oil, it is tainted and becomes unpalatable. Very small amounts of fuel oil (1 ppm) can contaminate drinking water in this way.

Because of these hazards, there are many regulations associated with the handling, transport, storage and use of fuels. These vary a great deal around the world, depending on the priorities of the authority imposing them. They are highly developed in several European countries, the USA, Canada and others, but negligible in a great number of Third World countries.

In the hotel industry, oil and gas are widely used for a variety of applications: for example to fire boilers for steam and hot water and for cooking. Emergency generators, using diesel oil, are provided in almost every hotel, and some hotels even have their own power plants. Fuel storage takes place in tanks either above or below ground. Fuel-oil storage tank sizes vary from less than 200 litre capacity up to large multiple tanks of up to 50 000 litres each.

When fuel oil or gas is used in hotels, risk of spillage and escape occurs at a number of points.

● Handling: spillage, overfilling, poor maintenance or repair on equipment carrying fuel, misuse of fuel for other purposes and intentional but inappropriate disposal into sewerage.
● Storage: tank corrosion, mechanical faults, installation mistakes in tanks piping or pumps, obsolete tanks left unattended.

Fuel storage practice for hotels

Good practice begins with a fuel storage policy, which aims to:

● Ensure that there is no environmental contamination resulting from current practice
● Ensure that future operations do not cause contamination
● Check that all storage facilities comply with local regulations
● Minimize fire risk
● Prevent economic losses from product leakage or from the high cost of contamination clean-up.

For the action plan:

1 Find out what fuel storage, handling and use regulations apply to you.
2 Identify national, regional and local fire and environmental regulatory requirements as well as house standards. Regulations vary greatly, and will have to be considered on an individual basis for each hotel.

Carry out a fuel storage inventory. Collect information on all fuel storage facilities, including data on the age of the tank, any cleaning, repairs or modifications in the past and the condition of the tank together with associated components such as piping and pumps. In addition, it is useful to record the throughput of the facility (through inventory records) and also to keep records of spills and leakage, preventative maintenance procedures and details of any official inspections. The facility should be evaluated in terms of its ability to control any potential leakage or spillage. Factors such as the provision of overspill protection, the local soil characteristics and the height of the tank above the local water table should be considered. Assess if agreed standards are being met.

Any possible leaks and spills should be identified (see Figure 6.5). One way of doing this is to look for any discrepancies in the inventory, such as a difference between delivery data and consumption data. Any apparent loss of inventory may be caused by a leak from a tank or its associated pipework. Visual inspection should be carried out on a regular basis by designated staff, looking for, and reporting on, evidence of spills such as staining of the ground, traces of oil in wells, dying vegetation and fuel odours, particularly in basement and ground-floor buildings. If leaks are suspected, tanks and pipework must be tested in order to identify the source.

Any spilt fuel will have to be removed and the contamination may require extensive and long-term corrective action, especially if the leak has existed for some time before detection, resulting in contamination of a large area. In this situation, extensive excavation may be required. Surface spills or leaks from pipework will usually be detected quickly and their spread can usually be limited by a prompt response.

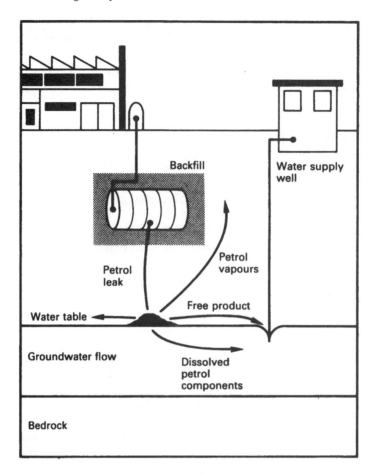

Figure 6.5   *Underground flow of leaking fuel and petroleum*

Any faulty tanks or pipework may need replacement and these should be designed to meet the latest regulations and recommendations. For example, tanks should have double walls of either steel or reinforced fibre glass. Metal installations, including tanks and pipework, require corrosion protection, with plastic cladding or cathodic protection by the installation of sacrificial anodes.

It may also be desirable to install leak-detection devices, such as automatic tank gauging, monitoring for vapours in the soil, interstitial monitoring and monitoring for traces of fuel oil in groundwater (see Figure 6.6).

Solid waste

Solid waste may be disposed of by incineration, composting, recycling, or burying in a landfill site. Of all of these alternatives, landfill is the most damaging to the environment and is also likely to become increasingly expensive as existing sites become full. The bulk of waste can be reduced by the use of a compactor at the hotel.

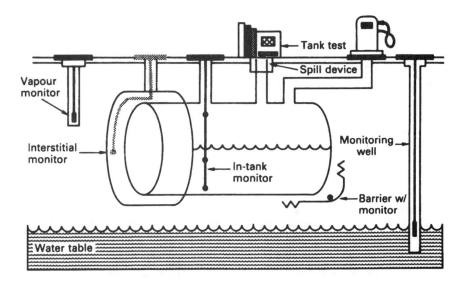

Figure 6.6 *Leak-detection alternatives*

Composting is suitable for waste which contains a high proportion of organic plant waste, such as trimmings from the kitchen, garden and ground waste, containers and packaging. Whether or not it is a suitable treatment depends upon the proportion of this organic material in the waste which relates to the operational characteristics of the hotel. The compost can be used on grounds and gardens as a dressing. This has the double advantage of reducing landfill and dependency on peat-based materials in the garden. Organic waste destined for the compost heap should be free of inorganic waste and substances toxic to plants. If the hotel produces large volumes of woody waste which cannot be converted into compost directly, these can be processed through a shredding machine and then composted to produce a 'forest bark' type of top dressing for use as a weed control on flower beds.

The possibility of incineration depends upon two factors. Does the waste have suitable characteristics for incineration in terms of its calorific value and moisture content and does it contain any chemicals which would release dangerous emissions? Second, is there an incineration plant close to the hotel, preferably one in which the combusted waste is used for district heating and/or electricity generation? Incineration by itself has the advantage of reducing landfill requirements, but it does produce CO_2 emissions.

Disposal of left-over chemicals

As most pesticides and herbicides are extremely toxic, due regard must be paid to this aspect when disposing of such wastes. The disposal methods selected will depend upon:

● The quantity of waste for disposal
● The chemical and biological degradability of the active ingredients

- Toxic properties of the active ingredients
- The concentration and physical form of the waste.

Before any wastes are disposed of, including empty containers, any instructions provided by the manufacturer or supplier should be consulted and followed. These instructions should be kept in a hazardous materials manual (see Chapter 5). Disposing of pesticides and herbicides to a landfill site is not generally an environmentally sound option. In most cases, wherever possible, incineration should be adopted, but at temperatures in excess of 1000°C and at a residence time of at least 2 seconds.

The following general guidelines should be followed:

- Do not re-use pesticide/herbicide containers
- Rinse out containers when empty and use the rinsing water for pest and weed control
- Puncture containers after they have been used to prevent their re-use
- Segregate pesticide/herbicide waste from general hotel waste
- Use an authorized waste-disposal contractor
- Use authorized disposal sites.

Recycling

The principles of recycling

Recycling achieves many goals:

1 It reduces waste materials.
2 It decreases the cost of waste disposal by reducing the volume of material to be collected, transported and dumped.
3 It reduces the requirement for landfill sites.
4 It maximizes the value of natural resources.
5 It often produces materials with a lower energy cost than the original product. For example recycling aluminium requires only 5 per cent of the amount required to extract aluminium from natural sources.
6 It often produces materials with a reduced amount of air pollution compared with the original production of that material.
7 Some recycled products have a cash value.

The first stage of recycling is to identify the types of materials that can be recycled and to consider how they can be segregated and collected ready for re-use or sale for reprocessing. Items are re-used when they are suitable for the same function or a comparable one. In reprocessing, the waste items are used to make the same or similar products (for example, the conversion of old newspapers into newsprint). One of the primary motivations to a sustainable recycling programme is the value of the waste to a recycling company or intermediary (Jaffe *et al.* 1993).

Recycling programmes are easily established but are fairly complex to run successfully. As a rule, the nearer to the origin of waste that recovery occurs, the less sorting and processing are needed before the material can be recycled. Another factor is that cleaner waste products usually obtain higher prices. Quality white paper has more value if it is segregated from newsprint. Coloured paper rapidly loses its value when it is mixed with other waste or refuse. Newspaper must be free of other waste products (particularly food and organic waste) in order to be effectively recycled.

One of the requirements of a recycling programme is the need for a storage area within the hotel and close to an access point where waste materials, which have been assembled for recycling, can be stored awaiting collection. There is a cost of recycling, in terms of the provision of labour to collect materials and space to store the materials. These costs must be balanced against the value of any payment received for the waste.

Common recyclable materials and resulting products

Aluminium

Although aluminium is one of the most abundant minerals found in the earth's crust, the processed metal has a high cost because of the large amounts of electricity required to separate the aluminium from its ore. Cans and foil are separated from steel cans using magnets, then shredded, delaquered and melted down to make various aluminium products. Aluminium cans often have a large scrap value because of the high energy cost associated with converting aluminium ores into the metal. Recycled material can be reprocessed at a much lower cost.

Steel cans (tins)

The protective tin coating on the steel is removed and the steel is then melted down to make many steel-based products. Tin cans make up only a small proportion of recycled steel and other sources include scrapped cars, appliances, farm machinery and industrial scrap.

Paper

Most high-grade virgin paper used in offices comes from trees. Therefore, anything which can be done to reduce the volume of purchases of high-quality paper will save money and also protect forests. However, paper is normally made from managed softwood plantations grown specifically for paper manufacture and so does not directly threaten rain-forests. In addition to the consumption of wood, the manufacture of paper also requires large volumes of water and energy. Because of the use of bleaching agents and other chemicals used in the manufacture of paper, the process can also produce polluted water discharges.

In principle, it should be possible to collect all waste paper for recycling, segregating high-quality papers (letter head, laser, photocopier) from lower-quality papers. Fine paper, which is collected from offices, has the highest resale value. After it is collected, it is cleaned, repulped and mixed with varying percentages of virgin pulp and made into paper to be used in the manufacture of boxboard, tissue, printing and writing papers, newsprint and liner-board.

Cardboard

This may need to be separated from paper and, in some countries, sorted into various grades. Another possible solution is to return the flattened, tied cardboard boxes to the supplier for re-use.

Glass

Waste glass usually has a low value, but this value can be maximized by segregating clear glass from coloured glass. Collected bottles and jars are crushed to make 'cullet'. This is mixed with sand, limestone and soda ash and melted in a high-temperature furnace. The molten glass is then moulded into new containers. Other current uses of waste glass include fibre glass, glass beads for reflective paint and some other building materials. Because of the abundance of the raw materials used in the manufacture of glass and the high energy cost involved in the conversion of waste, the economics of waste glass recycling is not always sound.

Plastic

Recycled plastics usually have a low value unless they can be carefully segregated into various types. Soft-drink bottles, milk bottles and laundry product bottles must be sorted into various groups of plastics before converting into pellets, which can be heated and moulded into desired shapes. Some of the products made from recycled plastics include drain-pipes, plastic bags, lids on non-food containers, insulation, flower boxes, clothes pegs, car bumpers, rope, carpet backing and household appliances. Shredded polyethylene terephthalate (PET) carbonated drink bottles can be used to produce a fibre for use in quilts, pillows, sleeping bags and coat linings. Where it is not possible to separate different types of plastic, mixed plastics can be shredded to produce 'plastic lumber'. Mixed plastic waste has a very low value.

Frying oil

Frying oil often has a commercial value and may be collected by brokers or manufacturers for a number of uses, including the base for cosmetics.

A fine-paper recycling programme

A paper processor or intermediary should be identified who is willing to collect (and possibly purchase) the waste before a start is made on the programme. Waste paper brokers will want to negotiate a price for the waste, which will depend on the level of service provided. For example, if the work of segregating and baling waste is done by hotel staff, this will increase the value of the waste, but may involve the hotel in additional costs.

Waste paper is generated in various areas of the hotel. A system of local collection is needed to allow the waste paper to be segregated from other office and general waste. A system of colour-coded bins may be used to make segregation of waste easy for staff. The segregated material will then need collection and transport to a central area where the materials can be sorted and packaged ready for collection by the processor.

The audit should reveal the quantity of paper waste broken down into a number of different categories. Surveys have indicated that it is possible, at best, to recover only 65 per cent of the potential waste paper because not all staff will support the scheme. The highest quality papers include computer paper and white bond, photocopier, laser paper and carbonless forms. If these papers can be kept separate from lower-quality papers they are likely to have a higher value. Mixed office paper including the above together with coloured paper has a lower value. The addition of other forms of paper, including newspapers and magazines, will have the lowest value. Discussions need to take place with local reprocessors in order to determine the most economical separation. The cost of separation must be balanced against the possible purchase price for the various categories of waste. Old newspapers and magazines can be recycled but the value is low and often it is not economic to recycle these materials, given the cost of transportation, storage and processing.

Most successful paper recycling programmes have four things in common:

1 A capable and enthusiastic coordinator to oversee all phases of the programme and planning through implementation to operation
2 A simple, reliable collection system, usually concentrated on one or perhaps two grades of paper
3 A secure market for the paper
4 An effective, on-going employee education and publicity programme.

Hotels vary in physical configuration and in the ways in which they operate, so it is unlikely that any one paper-recovery system will work for every location. Some of the companies involved in the collection of waste paper will provide assistance in setting up the system and may, for example, provide containers for the paper. The knowledge of a few underlying principles and techniques can simplify the task of designing an effective system for a hotel and of evaluating its feasibility. Whatever system is being considered, the proposal should be discussed with the housekeeping department, cleaning staff and any other hotel staff who are likely to be involved in operating the system.

First, high-volume, high-value paper should provide the focus for the system, with other grades of paper accommodated through subsidiary arrangements. Second, a suitable site must be found for the accumulation of the paper waste which is convenient for collection, such as the hotel loading or receiving dock. In addition to space for storage of the accumulated paper, a compactor and/or bailer may be required.

Separation of the paper waste may be done in one of two ways: by putting the responsibility for separation into the hands of either those who generate the waste or the cleaning staff. In the former scheme, each area which generates waste is provided with the requisite number of labelled containers for each category of waste. In the second scheme, all waste is put into single wastebaskets. Cleaners are provided with carts with two separate compartments, one for re-usable waste and one for other waste.

Case studies

Ramada Hotels and Renaissance Worldwide Hotels

Ramada Hotels and Renaissance Hotels use environmentally friendly guest amenities, e.g. soap, shampoo. Hotels in other regions are investigating similar product use. Some hotels in the group are purchasing dolphin-friendly tuna, while others use organically grown vegetables and herbs in their kitchens. The Ramada Caravelle in Frankfurt no longer serves individual, pre-packed portions of butter, cream, jam and condiments, and other members of the group in Germany use re-cycled toilet paper, tissues and printed materials for guests.

Forte plc

In 1980 when it was first suggested that aerosol cans were contributing to the depletion of the ozone layer, the Forte group carried out experiments which suggested that these should be banned on economic grounds as well. Forte found that the cost of the cans and propellent gas was over 60 per cent of the total cost of the product. Wastage occurred because it was easy to 'over-spray' and misdirect the spray, and up to 10 per cent of the usable product remained in the can, as the propellant gas was always exhausted before all the product could be extracted.

Economies have been achieved and product wastage reduced through purchasing cleaning materials in bulk containers and decanting them into re-usable containers and through using air-compressor, hand-operated spray pumps. This procedure also reduces the amount of containers and packaging thrown away.

Tamanaco Inter-Continental, Caracas

The Tamanaco Inter-Continental, Caracas, has started a programme of retrieving empty soap cases and returning them to the supplier for re-use. The intiative was instigated by one of the hotel employees in an attempt to save money and to reduce waste.

An agreement has been reached with the supplier and the hotel receives a rebate for each soap case it returns. The scheme was publicized within the hotel and the collection and counting of the cases are coordinated on a weekly basis by floor supervisors. The hotel management intends to give a token award to the floor that collects the greatest number of cases at the end of the year.

The programme began at the end of September 1992, since when 5000 empty soap cases have been collected. It is hoped that this will translate into an annual saving of $8400, and could lead to the recovery and re-use of other containers.

Sheraton Senggigi Beach Resort, Lombok

During the pre-opening stages of the Sheraton Senggigi Beach Resort in Lombok, Indonesia, in early 1991, the new chef was faced with a challenge. Supply and variety of local vegetables were extremely limited, and keeping up the high standards of cuisine expected would require a ready supply of fresh produce.

The two options for buying vegetables were importing from Bali, from where supply was unpredictable, or from Australia, which was costly. The chef set up an experiment with the local farming community, importing a wide range of vegetable seeds with the aim of seeing what would grow well in the local soil. Local people converted some ricefields into vegetable plantations, and the first experiments included chives, several kinds of lettuce, broccoli, Chinese and red cabbage, radishes, beets, celery, turnips, pumpkins and baby carrots. A wide range of herbs, including sage, dill, oregano, basil, parsley and lemon mint, were also planted.

All the produce from the project was purchased by the hotel, which meant for the farmers a diversification of crop production that was more profitable than the traditional rice crop. The hotel and farmers are experimenting with ways to increase yield and have introduced composting techniques, using waste from the hotel as a base. To sustain the development, the hotel and local farmers are studying methods to propagate seed stock from the 'new' vegetables, so that the community's development can be self-perpetuating.

The hotel has benefited, because it now has a regular supply of fresh vegetables, which were formerly imported from Australia at two or three times the cost. The local community has benefited, because it has a new source of income. The Sheraton group has benefited, because it has successfully defined an alternative approach to a problem that will enhance the image of the company on a national level. Sheraton's public image is particularly important, considering it plans to establish twenty hotels in Indonesia by the year 2000.

Inter-Continental, New Orleans

The Inter-Continental in New Orleans has been very successful in terms of efficient recycling and waste reduction. By sorting out room rubbish and food and beverage waste, it achieved a $79000 saving through the reduction of waste haulage, sale of recyclables and retrieving operating equipment.

Springs Hotel, Banff

The Canadian Pacific Hotels & Resorts' Banff Springs Hotel has implemented an extensive recycling programme for a whole range of products, resulting in a reduction in the volume of waste by seven-eighths. The hotel now only makes one daily trip to the garbage transfer station, saving on labour and fuel.

Programmes include:

- Collection and return of all refundable drinks containers
- Collection of non-refundable cans, which are shredded in the hotel's new recycling machine
- Collection and sorting of recycling paper before transportation to the recycling depot
- Collection and recycling of used grease from kitchen waste
- Collection and refining of motor oil from hotel vehicles
- Coathangers are collected from guest rooms and returned to the uniformed room for re-use. The uniform room now returns 1000 to 1500 coathangers to the laundry each month.

L'Hotel, Toronto

L'Hotel, Toronto (part of the Canadian Pacific Hotels & Resorts group) began its environmental programme in 1991 by initiating recycling programmes for paper, glass, cardboard, kitchen fats and scrap metal. Since then the hotel has found a number of simple, effective ways to reduce waste at source: guest room laundry bags (formerly plastic and disposable) are now made from retired bed sheets, and are re-usable. Fruit baskets and gifts are no longer wrapped in cellophane, and newspapers are no longer delivered in plastic bags. In January 1992 L'Hotel increased the scope of its recycling activities to include newspapers and cans (aluminum and steel). Changes in purchasing habits have also reduced waste that is harmful to the environment; environmentally friendly products such as ceramic mugs are used instead of styrofoam, and individual packaging has been discontinued in the staff restaurant. The programme has resulted in savings of $5000 on the hotel's waste disposal bill, $3450 on purchases of styrofoam cups and individual servings, $25 000 as a result of using energy-efficient lighting, $1900 received as a conservation rebate from the regional electricity company and savings of $8300 on the gas bill.

Maui Marriott Hotel

The Maui Marriott has obtained permission from its landowner to utilize an area of the property for composting the green waste that is generated in the hotel. The compost is used on the property for plant cultivation. This system has reduced the amount of waste that would normally end up in the hotel's compactor and/or the landfill.

A cardboard recycler has been installed at the Maui Marriott. The machine compacts the waste cardboard into bales, which can then be collected by a recycling agent. Since the introduction of the baler, the hotel has eliminated some of its landfill and waste removal costs.

On the same theme, recycling bins have been placed by all copy machines and in other strategic locations. Double-sided copying and the re-use of paper are encouraged. Green and white computer paper is collected separately from white paper and is taken periodically to a recycling centre.

Renaissance Hotel, Long Beach

At Long Beach, California, the Ramada Renaissance has attempted to reduce the amount of styrofoam it uses by offering guests a 25 per cent discount on coffee if they supply their own cups.

Le Meridien, Phuket

Le Meridien, Phuket, introduced the 3 Rs conservation programme in the hotel, encouraging staff to recycle, reduce and re-use. The plan was developed by hotel staff, and includes directives on topics as diverse as recycling light bulbs and using the on-site diving centre to remove underwater garbage from the sea. The programme was implemented through a detailed list of targets grouped under five themes: garbage treatment, waste, chemicals, product supplies and environment protection. These targets cover, for example, replacing plastic laundry bags with re-usable linen bags, recycling food as animal feed plus paper, glass, light bulbs, tin cans, batteries and candles, and avoiding CFC aerosols. All staff are encouraged to participate in the programme, as is the local community. The hotel newsletter and notices in bedrooms promote green tips and ask guests to help save energy and water. A free dive is offered at the diving centre for those who participate in the underwater cleaning programme.

Le Meridien, Newport Beach

Le Meridien, Newport Beach had its water boilers retrofitted to reduce contaminant emissions, and plans to install containment units in the chillers to prevent discharge of refrigerant from the purge system. In line with hotel concern to preserve air quality, 'bio-incubators' have been installed in grease traps and wet wells. These easy-to-install devices allow control and confinement of bacteria growth by accelerating biodegradation.

Hilton International

The Hilton International group is working on the systematic replacement of equipment, using CFC II with systems capable of operating with HCFC 123. These machines, although friendlier to the environment, are expensive. Where possible the group is also eliminating fossil-fuel use for heating purposes and replacing with natural gas.

Forte plc

All Forte establishments now use ozone-friendly aerosols and packaging materials and no longer use furniture filled with dangerous foam materials. Some 2000 company vehicles have been converted to running on lead-free petrol.

Semiramis Inter-Continental, Cairo

Engineers at Semiramis Inter-Continental in Cairo have invented a novel way of reducing the amount of harmful CFCs released into the atmosphere. Without access to a modern, freon-free refrigeration system (HCFC-123), the hotel must continue to use CFCs. Previously all refrigerant in the system was simply 'blown off' into the atmosphere before servicing or repairing the refrigerators. Similarly, air was removed from the pipes by purging them with extra refrigerant. Both practices release harmful CFCs into the atmosphere.

To combat this problem, the engineering department came up with a remarkable invention – a Freon Recovery Unit. Built by the engineers themselves, the unit consists of a hermetically sealed compressor joined to an air-cooled condenser coil. When applied to the refrigerator, the compressor gradually removes the freon from the system, condensing it so that it can be stored for re-use. This process removes virtually all the refrigerant in the refrigerator, reducing the CFC emissions, and saving money on the cost of repurchasing refrigerants.

Maui Marriott

The two main air-conditioning plants at the Maui Marriott use Refrigerant 11. In order to eliminate the constant venting of CFC into the atmosphere, the hotel has installed two oil-less purge systems. A reclamation system specifically for the air-conditioning plant has also been purchased – this removes all refrigerant from the two plants, stores it, cleans it and replaces it in the unit. If the refrigerant is contaminated beyond the system's ability to clean, recycle and purify it, the two Freon holding tanks can be shipped back to the factory, where the refrigerant can be reclaimed or disposed of.

The hotel is also researching the options of either replacing the two Freon air-conditioning plants with modern systems that can handle the new refrigerants or retrofitting the present systems to accommodate the same.

Summary

Environmental management represents a complex set of interacting issues which need to be considered in a holistic way. By considering all these factors as a system and how all the parts of the hotel system interact, it should be possible to make sensible decisions which allow the hotel to obtain the optimum benefit to the environment while not threatening the financial viability of the hotel. These interactions between aspects of the whole operation, from design, purchasing specification, production planning, stock management, waste management and waste disposal can provide financial as well as environmental benefits (see Figure 6.7). A hotel cannot afford to be altruistic, but by considering environmental management holistically it may be possible to invest savings made in one area into other activities which have less clear financial benefits.

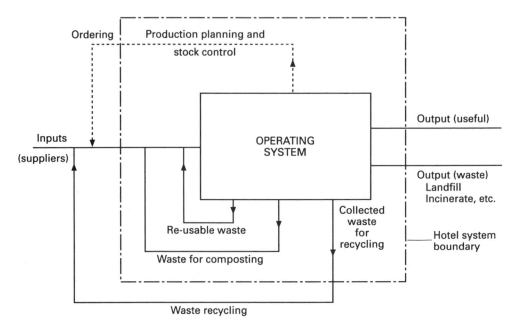

Figure 6.7   *The environmental management system*

References and further reading

Banks, G. H. and Collison, R. (1981). Food waste in catering. *Proceedings of the Institute of Food Science and Technology*, **14**, No. 4, 181–189.

Collison, R. Banks G. H. and Colwill, J. (1984). Food waste – its size and control. In Glew G. (Ed), *Advances in Catering Technology – 3*, London: Elsevier, pp. 157–166.

Collison, R. and Colwill, J. (1986). The analysis of food waste results and related attributes of restaurants and public houses. *J. Foodservice Systems*, **4**, 17–30.

Cummings, L. E. (1992). Hospitality solid waste minimisation: a global frame, *International Journal of Hospitality Management*, **11**, 3, 255–267.

Elkington, J., Knight, P. and Hailes, J. (1991). *The Green Business Guide*, Chapter 6, London, Gollancz.

Jaffe, W. F., Almanza, B. A. and Chen-Hua, J. M. (1993). Solid waste disposal: independent food service practices. *FIU Hospitality Review*, **11**, part 1, 69–77.

Lax, F. (1992). *Packaging and Ecology*, Leatherhead: PIRA.

Levy, G. M. (1993). *Packaging and the Environment*, London: Blackie.

Shanklin, C. W. (1993). Ecology age: implications for the hospitality and tourism industry. *Hospitality Research Journal*, **17**, part 1, 219–229.

Index

Page numbers in italic refer to captions to illustrations

Accor (UK) Management Ltd, case study, 25–6
Acid rain, 3, 9–10
Air conditioning:
  and air quality, 92–3
  and energy conservation, 64–8
Air pollutants, potential sources, 87–90
Air quality, 85–93
  costs, 91
  evaluation, 91
  heating and ventilation, 91–3
  improvement, 90–1
  sources of air pollutants, 87–90
Airborne micro-organisms, 90
Algae, toxic, 33, 34
Aluminium:
  recycling, 117
  in water, 37
Amstel Inter-continental, case study, 29
Aquifers, 33–4
Asbestos, 87–9
Atmosphere, 7–10
Audit:
  of energy use, 48, 53–4
  of equipment, 70
  of waste, 103–5, 107
  water use, 40–1

Bacteria, in water, 39, 40
Baths and showers, 43
Batteries, 106
Biological oxygen demand (BOD), 10
Boundaries of environment, 1–2
BS 5750/ISO 9000, 25
BS 7750, 25
Brundtland Report, 3, 4
Building Energy Management Systems (BEMS), 65, 66
Building materials, and pollutants, 87
Bulbs, light, 68
Business, and ecological processes, 17

Calorifiers, 36
Carbon dixoide, 3, 8, 90
Carbon monoxide, 90
Cardboard waste, disposal, 118
'Carrying capacity', 17–18
Case studies:
  energy management, 75–9
  environmental management, 25–30
  indoor environment, 100
  materials and waste management, 119–24
  water management, 44–5

Catalytic converters, 12, 106
CFCs (chlorofluorocarbons) 2, 6, 8, 106, 112
  control, 3
  displacers of ozone, 9
Change, faces of, 3–4
Chemical contaminants:
  table of concentrations permissible, 38
  of water, 37–9
Chemical energy, 48
Chemical hazards, 81–5
Chemical vapours, 87
Chemicals:
  handling, 85
  left-over, 115–16
Chemicals (Hazard Information and Packaging)
  Regulation (CHIP), 82
Chlorination, 39
Chlorofluorocarbons, see CFCs
Classification, Packaging and Labelling of
  Dangerous Substances Regulations 1984,
  82
Clean Air Act 1956, 3, 11
Cleaning, and water consumption, 69
Closed systems, 17
Cold water feed system, 35
Coliform bacteria, in water, 39
Colour temperature of light, 98–9
Combined heat and power (CHP), 51, 66–8
Combustion products, 87
Comfort, defined, 80–1
'Community Chest' scheme (Forte), 28–9
Composting, 115
Control of Substances Hazardous to Health
  Regulations (COSHH), 80, 82
Control of Pollution Act 1974, 3
Corrosive substances, 82
Crop yield, 6

Decibels, 93–4
Decision-making, 3
Deforestation, 8, 11
Degree days, 58–60
Deserts, 11
Detergents, waste, 107
Discharges, hazardous and harmful, 110
Disposal of unwanted materials, 6
District heating, 68
Domestic effluents, 10
Dust, 89, 90

Ecological labelling, 21
Ecological system, and business, 17
Ecology representatives (Novotel), 26
Ecosystems, 18
Education and research, 14, see also Training

Effluents, 10
Electrical energy, 48
Electromagnetic energy, 48
Electronic mail, 108
Electronic point of sale (EPOS) system, 108
Energy audit, 48, 53–4, *57*
Energy conservation measures, 60–3, *62*
   guest rooms 69
   guidelines by area, 64–72
   heating, ventilation and air conditioning, 64–8
   kitchens, 70–2
   lighting, 68–9
   monitoring and targeting, 63–4
Energy consumption, *55*
Energy conversions, and efficiency, 49–51
Energy Efficiency Office, 47, 54
Energy Managers, role, 53
Energy management, 47–79
   case studies, 75–9
   energy management programme, 53–74
   energy supplies, 48–53
   on equipment, 106
   principles, 47–8
Energy management programme, 53–74
   assessment of performance, 54–8
   degree days, 58–60
   energy audit, 48, 53–4
   energy conservation measures, 60–1
   energy recovery, 72–4
   monitoring and targeting, 63–4
   project appraisal, 60
Energy recovery, 72–4
Energy supplies:
   conversions and efficiency, 49–51
   forms of energy, 48
   non-renewable energy supplies, 51–3
   renewable energy supplies, 53
   secondary energy sources, 51
   units of energy, 49
Environment, defined, 1–3
Environmental awareness, 14
Environmental impact assessment (EIA), 22–5
Environmental issues, tourism and hospitality,
   13–14
Environmental management, 16–31
   case studies, 25–30
   environmental impact assessment, 22–5
   environmental policy, strategy and
      implementation, 18–21
   environmental system, 16–18
   system diagram, 124
Environmental management systems, complex,
   25
Environmental policy strategy and
   implementation, 18–21
Environmental pollution, 110–16
   diagram, 110
   and European Community, 3, 21
   gases, vapours and odours, 111–12
   left-over chemicals, 115–16
   solid waste, 114–15
   stored fuel, 112–14
   waste water, 111
Environmental Protection Act 1990, 21
Environmental purchasing, 106–7

Environmental system, 16–18
Equipment audit, kitchens, *70*
Ethical performance, 4
European Union/Community, 3, 21
Explosive substances, 82
Extraction fans, kitchens, 71

Fertilizers, 10
   in water, 32–3
Filtration of water, 39
Fish supplies, 6
Flammable substances, 82
Food:
   imported, 106
   preparation, 72
   waste, 18, 107–8
Forte plc, case studies:
   energy management, 75–6
   environmental management, 28–9
   materials waste, 120, 123
Fossil fuels, 5–6, 106
   and acid rain, 9
   heavily used, 51
Freon 11/Freon 12, 8
Freshwaters, 10
Friends of the Earth,
Fuel, 48
Fuel oils, 52
Fuel storage, 112–14
   inventory, 113
Furniture, disposal, 108

Gases:
   in air, 87
   waste, 111–12
Glass, recycling, 118
Global environmental issues, 7–12
Global warming, 7–9
Green Globe strategy (World Travel and Tourist
   Council), 14
Green policies for hotels, 14–15
Green political parties, 4
'Green Rooms' campaign, 28
Greenhouse effect, 7–9
Green Peace, 4
Groundwater, 33–4
Guest rooms, energy consumption, 69

Hardness in water, 37
Hawken, P., quoted, 16
Hayman Island Great Barrier Reef Resort:
   indoor environment, 100
   water case study, 45
Hazardous discharges, diagram, *110*
Hazardous materials, data sheet, 83, 116
Health and Safety at Work Act 1974, 80
Health and Safety at Work Regulations 1992,
   80
Heat and power system, combined, *67*
Heat exchangers, 73
Heat pumps, energy recovery, *74*
Heating, energy conservation, 64–8
Heating and ventilation, and air quality, 91–3
Herbicides, 84–5
   disposal, 116

Hertz, 94
Hilton International Group:
  indoor environment, 100
  materials and waste, case study, 123
  London Hilton, 45
Holiday Inn, Leicester, case study, 30
Hospitality, *see* Tourism
Hot water distribution, *36*
L'Hôtel, Toronto:
  energy conservation case study, 77
  materials and waste, case study, 122
Hotel Catering and Institutional Management
    Association, 14
Hotels:
  damage possibly done by, 13
  facilities affecting the environment, 13
  occupancy and water consumption, 42
  slow to act, 14
Hotels environment initiative, 14
Humidity:
  air quality, 90
  greenhouse effect, 8
  high, 87
Hyde Park Inter-continental, London, energy
    conservation, case study, 77–8

Incentives, for hotels, 14–15
Indoor environment, 80–101
  air quality, 85–93
  case studies, 100
  chemical hazards, 81–5
  light, 98–9
  noise, 93–8
  non-iodizing radiation, 99–100
Industrial effluents, 10
Industrial waste, 11
Infectious substances, 82
Initial environmental audit (IEA), 23, 24
Inputs, 5–6
  into environmental system, 16
Inter-continental Hotel Group case studies:
  energy management, 75, 78
  materials and waste, 121

Just-in-time techniques, 105

Kentucky Fried Chicken, case study, 29
Key-activated systems, 65–7
Key performance measures, 25
Kitchens, energy consumption, 70–2

Lakes, as water source, 34
Land, and pollution, 11
Landfill, 11, 102
Lasers, 99–100
Laundry, water management, 44–5
Lead, in water, 37
Leakage, fuel systems, 113–14, 115
Legionnaires' disease, 39, 40, 86, 90
Legislation, 21
Light, indoor environment, 98–9
Lighting, 68–9
London Hilton, water management, 45
Lumens, 98

Materials:
  case studies, 119–24
  environmental pollution, 110–16
  need, 102–3
  operations management, 107–9
  product purchasing, 105–7
  recycling, 116–19
  waste audit, 103–5
  and waste management, 102–25
Maui Marriott Hotel, case studies on materials
    and waste, 122, 124
Mechanical energy, 48
Le Meridien, San Diego, case studies:
  energy conservation, 77
  Newport Beach, 123
  Phuket, 122–3
Meridien Hotels, case studies, 29–30
Methane, 8, 87
Miami Dadeland Marriott, case studies,
    27–8
Micro-organisms, airborne, 90
Microwaves, 99
Mineral resources, depletion, 11
Mission statements, 14
Monitoring, post-audit, 23–4
Montreal Protocol 1987, 3

National Rivers Authority, 10, 21
Natural gas:
  increased use, 51–2
  supply, 52–3
Nature Conservancy, United States, 27
Nitrates, in water, 10, 32–3, 37
Nitrous oxide, 8
Noise, 93–8
  control, 97–8
  effects, 94–5, 112
  sources, 94
  tackling, 95–7
Non-ionizing radiation, 99–100
Non-renewable resources, 5–6, 21
  energy, 48, 51–3
Novotel, *see* Accor
Nuclear energy, 48, 53

Objectives of policy, 19
Oceans:
  and waste, 10–11
  as water supply, 34
Odours, 87, 111–12
Office areas, hazardous materials, 84
Oil:
  slicks, 11
  storage, 112
Open systems, 17
Operations management, 107–9
Organizational culture, and environmental
    policy, 18–20
Outputs, from system, 17
Ovens, energy saving, 72
Oxidation, biological, of water, 39
Ozone, 90
Ozone depletion, 6, 9

Packaging, minimizing, 105, 106

Paper:
    audit record, *109*
    recycling, 117, 118–19
    waste, 108
Partnerships on environmental policy, 20
Performance measures, 25
Periodic audits, 23
Pesticides, 6, 12, 37, 84–5, 91
    disposal, 116
Petroleum, fuel supply, 52
Photochemical smog, 11–12
Photocopier fumes, 84
Plastic, 118
Policy statements, 14, 20–1
'Pollution pays' principle, 111
Pollution, *see* Environmental pollution
Polychlorinated biphenyls (PCBs), 84
Potable and non-potable water supplies,
    34–5
Poverty, Third World, 12
Pressure groups, 4
Product purchasing, 105–7
Products, checklist on use and disposal, *104*
Purchasing policy:
    and the environmental, 105
    environmental purchasing, 106–7
    principles, 105–6

Radiation, non-ionizing, 99–100
Radioactive wastes, 12
Radon gas, 87
Rain, increased acidity, 3, 9–10
Rainforests, 11
Ramada Group, case studies:
    energy management, 75, 78
    environmental management, 27
    indoor environment, 100
    materials and waste, 119–20
    water, 44
Recycled materials, 106
Recycling, 103, 105, 116–19
    aluminium, 117
    glass, 118
    paper, 117–18, 119
    plastic, 118
    steel cans, 117
Refrigeration, energy conservation, 72
Regency Inter-continental, Bahrain, case studies:
    energy conservation, 76
    water management, 44–5
Regulation of environment, 14
Renaissance Hotels, case studies, materials and
    waste, 119–20, 122
Renewable energy supplies, 53
Renewable resources, 6
Re-use of material, 108
Ritz Hotel, energy consumption, 65
Rivers:
    and pollution, 10
    water source, 34
Road traffic, 12
Royal Orchid Sheraton Hotel and Towers,
    Bangladesh, water management, 45, 111
Sanitation, kitchens, 72
Sea, *see* Oceans

Secondary energy sources, 51
Semiramis Inter-continental, Cairo, case studies on
    materials and waste, 123
Sewage treatment, 111
Sheraton Hotels, case studies:
    energy management, 79
    materials and waste, 120–1
Sick Building Syndrome (SBS), 86
Smog, 11–12
Smoke haze, 11
Smoking, 87
Software for hazardous materials recording, 82
Soil erosion, 11
Solar energy, 6
Solar power, 53
Solar radiation, dispersal, 7–8
Solid waste, 114–15
Solvents, a danger, 84
Springs Hotel, Banff, case studies, materials and
    waste, 121
Staff, planning environmental management, 19
Steel cans, recycling, 117
Stored fuel, 112–14
Sulphur dioxide, 3
Sun, 17
Sunbeds, 99, 100
Surface water, 34
Sustainable development, 4–5, 21
Swiss Hotel Association, case studies, 28
Systems, closed and open, 17

Tamanaco Inter-continental, Caracas, case studies,
    30
    materials and waste, 120
Targeting, post-audit, 23
Targets, 14
Temperature:
    control, 81
    increase, 8
    urban, 11
    water in hotels, 36, 42–3
Thermal energy (heat), 48
Thermal wheels, 73
Thermostatic radiator valves (TVR), 65, 66
Tins, recycling, 117
Tourism and hospitality, and environment,
    12–15
Toxic substances, 82
Training:
    energy saving, 48, 64
    and water wastage, 42
Typewriting correction fluid, 84

Ultra-violet radiation, 9, 100
United Nations, Rio de Janeiro Earth Summit
    Conference 1992, 3
Units of energy, 49
Urban effects, 11–12

Vapours, waste, 111–12
Vehicle emissions, 91, 106
Ventilation, 86, 91–3
    and energy conservation, 64–8
    types, 92
Vibration, source of noise, 97

Waste, 17
Waste audit, 103–5, 107
Waste disposal, 6
Waste management, need for, 102–3
Waste minimization, 102
Waste water, 37, 111
Water:
  acidification, 10
  action plan on quality, 39–40
  and chemicals, 37–9
  contamination, 32–3
  defects and treatment, 40
  and environment, 32–3
  feed system, 35, 36
  hardness, 37
  in hotel, 35–7
  impurities, 37
  nitrates, 10
  quality, 34–5, 37–40
  supplies, 33–7
  treatment, 39
  waste water, 37, 111
Water Act 1989, 10
Water consumption:
  assessing current performance, 40–1
  benefits, 43–4
  control, 40–4
  reducation of wastage, 42–3
Water management, 32–46
  case studies, 44–5
  control of consumption, 40–4
  improvement of water quality, 37–40
  water and environment, 32–3
  water supplies, 33–7
Water softeners, 37–8
Water use audit, 40–1
Water vapour, 87
Wells, 33–4, 40
White Paper 1994, *Sustainable Development: the UK Strategy*, 21
  1990, *This Common Inheritance*, 3
Wooden products, 106
World Health Organization, 86
  and water quality, 38, 39
World Travel and Tourism Council (WTTC), 14
World Travel and Tourism Environmental Review, criteria, 13–14

Young, Steven, *The Politics of the Environment*, 4

Further reading from Butterworth-Heinemann

The Development and Management of Visitor Attractions

John Swarbrooke
Senior Lecturer, Centre for Tourism, Sheffield Hallam University

This book fills a major gap in the literature on this rapidly expanding and crucial area of tourism. It reflects the latest developments in the field and anticipates the impact of political, economic, social and technological change on visitor attractions.

0 7506 19791   1995   PAPERBACK

Global Tourism

Edited by William Theobald
Professor and Director, Travel and Tourism Program, Purdue University

Using the perspective and expertise of twenty-nine contributors, *Global Tourism* draws together current thinking and practice in the industry.

The purpose is to allow readers to examine critical issues and problems facing the tourism industry. The problems are complex and interwoven and they suggest a variety of crises such as overcrowding of tourist attractions; resident-host conflicts; loss of cultural heritage; inflation and escalating land costs; and a host of other political, sociocultural and economic problems.

0 7506 2353 5   1995   NEW IN PAPERBACK

Dictionary of Travel, Tourism and Hospitality

S Medlik
Former Head of Department of Hotel, Catering and Tourism Management, University of Surrey

Defines more than 1000 terms used in the study of travel, tourism and hospitality and explains the meaning of a similar number of abbreviations.

It also describes 500 British and international organizations and lists key data for 200 countries. The total of some 3000 entries represents a major source of information and a unique source of reference.

0 7506 0953 2   1993   PAPERBACK

For ordering details or further information about these, or any of Butterworth-Heinemann's books, please call 01865 314460 quoting code B509BSD

The Economics of Leisure and Tourism

John Tribe

Senior Lecturer in Economics, Buckinghamshire College, a college of Brunel University

This title, which assumes no prior knowledge of economics, applies and sets economics in the context of the leisure and tourism industries. There is an emphasis on economics for management and marketing and an economic analysis of current issues e.g. environmental issues and sustainable tourism.

Containing chapter objectives, chapter summaries, review questions and case studies, the book is intended to give students an accessible introduction.

The Economics of Leisure and Tourism is a new and fresh text which tackles issues that will affect the sector into the next century including globalization, virtual leisure and the grey and green revolutions.

Case studies include:

❶ *The National Lottery*
❷ *Port Aventura (Spain)*
❸ *BA's Global Strategy*
❹ *Holiday Inn International*
❺ *Virgin Radio*
❻ *Eurotunnel*
❼ *Business-class travel*
❽ *Video-on-demand*
❾ *Disneyland, Paris*
❿ *CCT in action*

Issues tackled include:

❶ *What was the BSkyB floatation*
❷ *What's the economic function of ticket touts?*
❸ *Can you haggle at the Hilton?*
❹ *Are CD prices a rip off?*
❺ *Should we privatize the BBC?*
❻ *What is the green revolution?*
❼ *Can tourism save Castro's Cuba?*
❽ *Why do we pump raw sewage into the sea?*
❾ *Why do we subsidize opera but not rock?*
❿ *Will leisure disappear into cyber-space?*

0 7506 2342 X   1995   PAPERBACK